DHEA and the Brain

Nutrition, brain and behaviour

Edited by Chandan Prasad, PhD
Professor and Vice Chairman (Research)
Department of Medicine
LSU Health Sciences Center
New Orleans, LA, USA

DHEA and the Brain

Edited by

Robert Morfin

Laboratoire de Biotechnologie
Conservatoire National des Arts et Métiers
Paris, France

CRC Press
Taylor & Francis Group
Boca Raton London New York

CRC Press is an imprint of the
Taylor & Francis Group, an **informa** business

A TAYLOR & FRANCIS BOOK

CRC Press
Taylor & Francis Group
6000 Broken Sound Parkway NW, Suite 300
Boca Raton, FL 33487-2742

First issued in paperback 2020

© 2002 by Taylor & Francis Group, LLC
CRC Press is an imprint of Taylor & Francis Group, an Informa business

No claim to original U.S. Government works

ISBN-13: 978-0-367-45495-1 (pbk)
ISBN-13: 978-0-415-27585-9 (hbk)

Visit the Taylor & Francis Web site at
http://www.taylorandfrancis.com

and the CRC Press Web site at
http://www.crcpress.com

Typeset in Baskerville by
Integra Software Services Pvt. Ltd, Pondicherry, India

British Library Cataloguing in Publication Data
A catalogue record for this book is available
from the British Library

Library of Congress Cataloging in Publication Data
A catalog record has been requested

Contents

Contributors

Kazutaka Aoki
The Third Department of Internal
 Medicine
Yokohama City University School of
 Medicine
3–9 Fuku-ura Kanazawa-ku
Yokohama 236-0004
Japan

Maria Luisa Barbaccia
Department of Neuroscience
University of Rome "Tor Vergata"
Edificio F-sud- Via Tor Vergata
00133-Rome-Italy

Rachel C. Brown
Interdisciplinary Program
 in Neuroscience
Georgetown University School
 of Medicine
3900 Reservoir Rd. NW
Washington, DC 20007

Richard Hampl
Institute of Endocrinology
2 Národní, 116 94 Praha
Czech Republic

J.J. Hauw
Service de Neuropathologie
Raymond Escourolle
Groupe Hospitalier Pitié-Slapétrière
47–83 Boulevard de l'Hôpital
75651 Paris Cedex 13
France

J. Herbert
Department of Anatomy
Downing Street
University of Cambridge
Cambridge CB2 3DY
UK

Martin Hill
Institute of Endocrinology
2 Národní, 116 94 Praha
Czech Republic

Clemens Kirschbaum
Institute of Experimentell Psychology
II University of Duesseldorf
Geb 23.02 Ebene 01, Raum 44
Universitaetsstrasse 1 D-40225
Duesseldorf
Germany

Ying Liu
Division of Hormone Research
Departments of Cell Biology
Georgetown University School
 of Medicine
3900 Reservoir Rd NW
Washington, DC 20007

J. Loeper
INSERM U 538
Faculté de Médecine
Saint Antoine
27 rue de Chaligny
75571 Paris
Cedex 12

Maria Dorota Majewska
Division of Treatment Research
  and Development
National Institute on Drug Abuse
NIH, 6001 Executive Blvd Rm.
4123, MSC 9551, Bethesda
MD 20892-9551, USA

Doris Mayer
Research Group "Hormones and
  Signal Transduction"
Deutsches Krebsforschungszentrum
Im Neuenheimer Feld 280
69120 Heidelberg
Germany

Synthia H. Mellon
Department of Obstetrics Gynecology
  & Reproductive Sciences
The Center for Reproductive Sciences
  and The Metabolic Research Unit
University of California
513 Parnassus Ave. Room 1661 HSE
Box 0556, San Francisco
CA 94143–0556, CA, USA

R. Morfin
Laboratoire de Biotechnologie
Conservatoire National des
Arts et Métiers, 2 rue Conté
75003 Paris
France

L.-Y. Ngai
Department of Anatomy
Downing Street
University of Cambridge
Cambridge CB2 3DY,
UK

Vassilios Papadopoulos
Division of Hormone Research
Departments of Cell Biology
Pharmacology and Neuroscience
Interdisciplinary Program in
  Neuroscience
Georgetown University School of
  Medicine
3900 Reservoir Rd. NW
Washington, DC 20007

D. Pompon
Laboratoire d'Ingéniérie de
Protéines Membranaires
Centre de Génétique Moléculaire
Centre National de la
Recherche Scientifique
91190 Gif sur Yvette
France

Chandan Prasad
Section of Endocrinology
Department of Medicine
LSU Health Sciences Center
1542 Tulane Avenue
New Orleans LA 70112
USA

Victor I. Reus
Department of Psychiatry
School of Medicine
University of California
San Francisco
Langley Porter Psychiatric Institute
San Francisco
California 94143-0984
USA

Hisahiko Sekihara
The Third Department of Internal
  Medicine
Yokohama City University School of
  Medicine
3–9 Fuku-ura Kanazawa-ku
Yokohama 236-0004
Japan

Tatiana Sidiropoulou
Department of Neuroscience
University of Rome "Tor Vergata"
Edificio F-sud- Via Tor Vergata
00133-Rome-Italy

Luboslav Stárka
Institute of Endocrinology
2 Národní, 116 94 Praha
Czech Republic

Frank Svec
Section of Endocrinology, Louisiana
State University Medical School
1542 Tulane Ave, New Orleans
LA, 70112, USA

M. Trincal
Laboratoire de Biotechnologie
Conservatoire National des
Arts et Métiers, 2 rue Conté
75003 Paris
France

Oliver T. Wolf
Institute of Experimentell Psychology
II, University of Duesseldorf Geb
23.02, Ebene 01, Raum 44
Universitaetsstrasse 1, D-40225
Duesseldorf
Germany

Owen M. Wolkowitz
Department of Psychiatry
School of Medicine
University of California
San Francisco
Langley Porter Psychiatric Institute
San Francisco
California 94143-0984,
USA

Series preface

Modern medical science has made impressive progress in understanding the role of dietary nutrients – both macro- and micronutrients – in the maintenance of normal health and in the prevention of diseases like scurvy, pellagra, marasmus, Kwashiorkor and many others. We also have gained a better understanding of fetal and infant nutrition. Despite these obvious successes, we have failed to fully realize the potential of certain dietary nutrients and supplements in the control of bodily functions such as mental performance. These dietary components include not only macronutrients (e.g. protein, carbohydrate and fat) but micronutrients (e.g. amino acids, vitamins and minerals) as well as metabolites (e.g. melatonin and dehydroepiandrosterone (DHEA)).

Contemporary interest in maintaining and enhancing both body and mind through diet and dietary supplementation has generated a multibillion dollar industry throughout the world. The use of diet to enhance mental function is not a recent phenomenon. For example, in the medieval holistic view of nature, mood was thought to be modulated by foods.

The heightened public and scientific interest in nutrition and brain function underscores the need for a regular authoritative overview of this subject. Currently, no publication addresses this need. Nutritional Neuroscience, a journal recently launched by Harwood Academic Publishers, is limited to specialized scientific investigations.

The Nutrition, Brain, and Behavior Series will provide a forum in which basic scientists and clinicians can share knowledge and perspectives regarding the role of dietary nutrients and supplements in nervous-system functions at structural, chemical, and behavioral level. The first volume in this series is *DHEA and the Brain* edited by Professor Robert Morfin of the University of Paris, France. Subsequent issues will focus on antioxidant vitamins, essential fatty acids, amino acids (tyrosine and tryptophan), nutrition and developing brain, and nutrition and the adult brain. I hope this series will become an essential source book for nutritionists, psychiatrists and internists.

I would like to take this opportunity to thank Professor Robert Morfin, the editor for the first volume in the series and the members of the Editorial Advisory Board (Drs Galler, Guedes, Huether, Kastin and Lieberman) for their help in getting this project off the ground.

Chandan Prasad, PhD
Series Editor

Preface

Since the discovery of DHEA and of its age-related decrease, many investigations have been aimed at defining its effects and its possible mechanism of action. Recently, many beneficial effects have been described after DHEA administration to normal, aged or diseased animals and humans. These effects include anti-cancer, anti-osteoporosis, anti-atherosclerosis, anti-diabetes, immuno-stimulating and memory enhancing activities that have led DHEA to be made available openly in the market place with strong concerns expressed by some scientists and physicians. It has been suggested that prolonged self-administration of DHEA may result in a number of health problems in the future due to its unknown mode of action and the production of steroid derivatives in most of the organs flooded by the drug. Most of the investigations devoted to DHEA metabolism and to its mode of action were either not conclusive or could not be reproduced. The reported finding of a specific binding of DHEA by an unknown protein species has not been confirmed and DHEA transformation products and yields vary considerably from species to species and from one to another tissue or organ. At present, and even with the multiple effects described, no work is available in support of a possible mechanism of DHEA action. DHEA production in the brain has resulted in it being termed a "neurosteroid", but its production from the pregnenolone precursor by the classical 17α-hydroxylase, $C_{17,20}$-lyase system could not be proved.

It, therefore, becomes obvious that all well-known paradigms cannot be applied with success to the DHEA story, and that new concepts need to be found for deciphering its production and its mode of action. Selected investigators have been faced with this problem, and their chapters in this book are presented in such a way as to bring out new ideas or new approaches devoted to investigations with DHEA.

DHEA administration to animals and humans results in extensively studied effects including well being, improved cognition and some decrease in age related onset of pathologies. Such works are extensively reviewed in this book in chapters from Frank Svec, Oliver T. Wolf and Clemens Kirschbaum, Victor Reus and Owen Wolkowitz, Lai-Yeung Ngai and Joe Herbert, Kazutaka Aoki and Hisahiko Sekihara, and Maria Dorota Majewska. Then, because DHEA administration to an organism involves any tissue and organ as a possible target where DHEA may be either directly active or transformed into more potent metabolites, the first question is what is DHEA metabolism in tissues and organs, including brain. This point is addressed in the chapters by Richard Hampl *et al.* and Mathieu Trincal *et al.*

Because liver is the first organ to be flooded with DHEA after *per os* administration, Doris Mayer's chapter gives clues to the second question concerned with DHEA effects in liver and liver produced DHEA metabolites which may be of importance for the brain.

The third key question is how DHEA is made in the brain. Production of the pregnenolone precursor by oligodendrocytes has been demonstrated, but many investigations failed to prove pregnenolone conversion into DHEA through the classical enzymatic pathway. These facts are reviewed in chapters by Maria Luisa Barbaccia and Tatiana Sidiropoulou, and by Synthia H. Mellon. In addition, new findings on brain DHEA production through a reactive oxygen species mediated mechanism are shown in the chapter by Rachel C. Brown *et al.*

The fourth question is related to the neuroprotective effects of DHEA. Because such effects may result either directly from DHEA or from one of its metabolites, answers to the previous questions were needed and new perspectives are presented by Richard Hampl *et al.* and Mathieu Trincal *et al.*

The fifth question is related to DHEA mechanism of action in the brain. Possible cross talk with neuropeptides and antiglucocorticoid effects of DHEA and of its metabolites are developed in chapters by Chandan Prasad and by Mathieu Trincal *et al.*

From these chapters, the reader may nurture new ideas and protocols that should contribute to resolving the DHEA mysteries in the brain. One of the possible leads could be that DHEA is the described "mother steroid", and as such, may be the precursor for metabolites whose activity and mode of action do not follow established rules for steroid hormones.

Robert Morfin

Chapter 1

The brain, a putative source and target of DHEA

Maria Luisa Barbaccia and Tatiana Sidiropoulou

NEUROSTEROID: THE CONCEPT AND ITS IMPLICATIONS

The term "neurosteroid" (Corpechot *et al.*, 1981) was introduced after it was observed that, even weeks after the surgical removal of peripheral steroidogenic tissues, relevant concentrations of dehydroepiandrosterone (DHEA) and DHEA-sulfate (DHEAS), as well as of pregnenolone and pregnenolone sulfate were present in the brain of adult rat, a species in which, at variance with humans, the adrenal DHEA production appears confined to early developmental stages. More-over, in rodents and non-human primates adrenalectomy or dexamethasone treatment reduce the concentrations of DHEA and other neurosteroids by a lower extent in brain than in plasma (Corpechot *et al.*, 1981; Robel and Baulieu, 1995; Roscetti *et al.*, 1998). It was therefore suggested that neurosteroids may be synthes-ized in brain and that this process was not under the control of the pituitary factors that control peripheral steroidogenesis.

Neural cells express the mitochondrial benzodiazepine receptor and the peptide diazepam binding inhibitor (DBI), its endogenous ligand, as well as most of the enzymes, cytochrome P450scc, 3β-hydroxysteroid-dehydrogenase, 17β-hydroxy-steroid dehydrogenase, 5α-reductase and 3α-hydroxysteroid-oxidoreductase (Besman *et al.*, 1989; Karavolas and Hodges, 1991; Robel and Baulieu, 1995; Celotti *et al.*, 1992; Mellon and Deschepper, 1993; Costa *et al.*, 1994; Mensah-Nyagan *et al.*, 1999) which enable *de novo* neurosteroid synthesis, and the metabo-lism of circulating steroid precursors in the brain. While there is convincing evidence for the competence of neural cells to synthesize *de novo* pregnenolone from choles-terol, as well as other neurosteroids, i.e., progesterone and its 3α-hydroxylated-5α-reduced metabolites (Le Goascogne *et al.*, 1987; Jung-Testas *et al.*, 1989; Karavolas and Hodges, 1991; Robel *et al.*, 1991; Guarneri *et al.*, 1992; Papadopoulos, 1993; Costa *et al.*, 1994; Mensah-Nyagan *et al.*, 1999; Serra *et al.*, 1999), the pres-ence and nature of the biosynthethic pathway(s) leading to DHEA synthesis in brain are still a matter of controversy (Baulieu and Robel, 1998). Several groups attempted to identify in adult mammalian brain the bifunctional enzyme cytochrome P450,17-hydroxylase/17,20-lyase (P450 c17), which catalizes the production of DHEA from pregnenolone in peripheral steroidogenic tissues. Cytochrome P450 c17 messenger RNA is expressed by various CNS structures of rat embryos (Compagnone *et al.*, 1995), but conflicting results have been reported in adult rodent brain, where its presence was observed by some (Stromstedt and Waterman, 1995), but not by other investigators (Mellon and Deschepper, 1993). Also, P450 c17 activity and immunoreactivity failed to be detected in adult rodent brain (Le Goascogne *et al.*, 1991; Robel *et al.*, 1991). Recently, by means of RT-PCR/Southern blot analysis it has been reported that astrocytes and neurons, but not oligodendrocytes, from neonatal rat brain express the messenger RNA encoding for P450 c17 (Zwayn and Yen, 1999). These authors also showed that astrocytes, and to a much lower extent neurons, are able to convert exogenously supplied pregnenolone to DHEA, though with a rather low yield. These controversial findings led to investigation of the existence of alternative enzymatic and non-enzymatic biosynthetic pathways for DHEA synthesis in neural cells (Lieberman, 1995; Cascio *et al.*, 1998; Brown *et al.*, 2000).

The interest in neurosteroids was further boosted by the demonstration that they exert pharmacological effects mediated by mechanism(s) of action which differ from those responsible for the effects of classical steroid hormones (i.e., glucocorticoid, mineralocorticoids, androgens, estrogens and progestins). The latter, following their binding to specific cytosolic receptors, form hormone–receptor complexes that, after translocating to the nucleus, either regulate target gene promoters or interact with transcription factors to modulate gene expression (Evans, 1988). Although some neurosteroids may act at the genomic level (Koenig *et al.*, 1995; McEwen, 1991; Rupprecht and Holsboer, 1999), they also interact with specific membrane sites and modulate, in a selective and stereospecific manner, the function of ligand-gated ionotropic receptors (Majewska *et al.*, 1986; Puia *et al.*, 1990; Paul and Purdy, 1992; Lambert *et al.*, 1995; El-Etr *et al.*, 1998; Park-Chung *et al.*, 1999; Rupprecht and Holsboer, 1999). To lay emphasis on this novel non-genomic mechanism of action, that implies a function for neurosteroids as direct modulators of intercellular signaling in brain, Paul and Purdy (1992) introduced the term "neuroactive steroid".

Limiting our attention to DHEA and DHEAS, these two "neuroactive neurosteroids" have been shown to positively modulate the NMDA-type of glutamate receptors, either directly or indirectly, through an interaction with the σ receptor (Monnet *et al.*, 1995; Maurice *et al.*, 1998); DHEAS is also a negative modulator of the action of GABA at GABA$_A$ receptors (Majewska, 1995; Randall *et al.*, 1995; Meyer *et al.*, 1999), an effect that, according to some (Imamura and Prasad, 1998) but not other authors (Sousa and Ticku, 1997), is exhibited also by DHEA. On the other hand, androsterone and 5α-androstane-3α,17β-diol/3α-ADIOL, 3α-hydroxylated, 5α-reduced metabolites of DHEA, are positive allosteric modulators of the action of GABA at GABA$_A$ receptors (Gee *et al.*, 1988; Morrow *et al.*, 1990; Lambert *et al.*, 1995; Frye *et al.*, 1996), though showing *in vitro* a lower potency with respect to the 3α-hydroxy-5α-pregnane steroids, i.e., allopregnanolone and allotetrahydrodeoxycorticosterone. Thus, DHEA and DHEAS have been reported to exert diversified pharmacological actions in the central nervous system, either *in vitro* or *in vivo*: (i) enhancement of neuronal survival and astrocyte differentiation (Bologa *et al.*, 1987); (ii) stimulation of Ca^{++} influx in primary cultures of mouse embryonic neocortical neurons, leading to neurite extension and appearance of varicosities (Compagnone and Mellon, 1998); (iii) increase of NMDA-mediated long term potentiation and population spike amplitude in hippocampus *in vitro* (Randall *et al.*, 1995; Meyer *et al.*, 1999); (iv) potentiation of neuronal NMDA-mediated responses in hippocampus *in vivo* (Bergeron *et al.*, 1996; Debonnel *et al.*, 1996); (v) potentiation of NMDA-induced norepinephrine release in hippocampus (Monnet *et al.*, 1995); (vi) pro-mnestic effects after intracerebral injection to mice (Flood *et al.*, 1988); (vii) anxiolytic action (Melchior and Ritzman, 1994; Prasad *et al.*, 1997) and increase of pentobarbital and ethanol-induced sleeping time (Melchior and Ritzman, 1992), after systemic administration.

Altogether, this evidence suggests that endogenous DHEA/DHEAS may play a role in brain pathophysiology and, similar to other neurosteroids, may modulate intercellular signaling in brain. This modulatory role, however, highlights a crucial and, as yet, poorly studied aspect concerning "whether" and "how" neurosteroidogenesis is regulated. Based on studies in animals, several, and not necessarily mutually exclusive, possible mechanisms may be operative: (i) the brain neurosteroid concentrations may change in parallel with increases or decreases of neuroactive steroids or of circulating precursors, as brain cells express the enzymes involved in the metabolism of the latter; (ii) local neurosteroid production may be affected by brain regional differences or development-related changes in enzyme expression and/or activity; (iii) neurosteroidogenesis may be modulated by intracellular second messenger pathways activated by specific extracellular signals. The latter, involving a cross-talk between neurosteroidogenesis and neuronal activity, would provide distinct neuronal circuitries with a mechanism to rapidly tune the local neurosteroid repertoire. We have investigated this possibility by taking advantage of an "*in vitro*" model system consisting of brain tissue minces ($300 \times 300\,\mu m$) from adult male rat. This model, maintaining to an acceptable extent the integrity of the cellular elements and of their reciprocal connections, allowed us to study the modulation of neurosteroidogenesis minimizing the interference of blood-borne steroids. The steroids, extracted from brain tissue homogenate in ethyl-acetate and purified by HPLC were measured by radioimmunoassays, carried on with specific antisera (Roscetti *et al.*, 1998). In this chapter, we shall focus our attention on DHEA and report on results showing that in adult rat brain tissue the synthesis/accumulation of endogenous DHEA can be modulated in a time frame of minutes by ascorbic acid, with a mechanism involving adenylate cyclase-coupled serotonin receptors. Moreover, we shall present evidence indicating that changes in the endogenous concentrations of DHEA and one of its metabolites, 5α-androstane-3α,17β-diol, correlate with specific mood alterations in humans.

REGULATION OF PREGNENOLONE AND DHEA, SYNTHESIS/ACCUMULATION IN RAT BRAIN TISSUE

Earlier studies in our laboratory showed that the enhancement of intracellular cyclic AMP levels was associated with an increase in the synthesis/accumulation of pregnenolone and progesterone in brain cortical minces prepared from adult male rats (Barbaccia *et al.*, 1992). The action of dibutyryl-cyclic AMP was mimicked by isobutylmethylxanthine, an inhibitor of cyclic AMP phosphodiesterase, and by the adenylate cyclase activator forskolin, but not by its inactive analog 1,9-dideoxy-forskolin. This effect peaked between 15 and 30 minutes and vanished by 60 minutes. A stimulatory action of cyclic AMP on neurosteroidogenesis had been previously reported in glial cells (Hu *et al.*, 1987); in those experiments, however, the effect was evaluated after 24–48 hours and was associated with an increased expression of steroidogenic enzymes. A pretreatment of the brain cortical minces with the protein synthesis inhibitor cycloheximide or with PK 11195, a partial agonist of the mitochondrial benzodiazepine/diazepam binding inhibitor (DBI) receptor, inhibited the cyclic AMP-induced increase of pregnenolone (Roscetti *et al.*, 1994). These results were interpreted as to indicate that cyclic AMP may increase neurosteroid synthesis/accumulation by transiently enhancing the availability of the peptide diazepam binding inhibitor or of its processing products, which act as endogenous agonists of the mitochondrial benzodiazepine/DBI receptor and increase the synthesis of pregnenolone, both in brain and peripheral steroidogenic cells (Besman *et al.*, 1989; Mukhin *et al.*, 1989; Papadopoulos, 1993; Costa *et al.*, 1994). This view is supported by the observation that ACTH, by stimulating intracellular cyclic AMP concentrations, mediates an increase in DBI processing towards peptide fragments active at the mitochondrial benzodiazepine/DBI receptor in rat adrenlas (Cavallaro *et al.*, 1992). These results also prompted us to evaluate whether the activation of specific neurotransmitter receptors positively coupled to adenylate cyclase (β-adrenergic, dopaminergic, serotoninergic) would modulate the synthesis/accumulation of neurosteroids in rat brain tissue.

Ascorbic acid stimulates pregnenolone and DHEA synthesis/accumulation in rat brain cortical tissue

The exposure of brain cortical minces to various concentrations of serotonin (0.4–50 μM) dopamine (1–50 μM) and isoproterenol (1–10 μM), revealed that serotonin (10 μM), but not dopamine nor isoproterenol, increased by about 40% pregnenolone concentrations (Roscetti *et al.*, 1998). Unexpectedly, the addition of ascorbic acid (0.01%) to the incubation buffer, in order to prevent monoamine oxidation, enhanced the levels of pregnenolone also in "control" samples. Ascorbic acid is a vitamin for humans and other species (which carry a nonfunctional gene of the enzyme L-gulono-γ-lactone oxidase, required for the synthesis of ascorbic acid from L-glucose) but not for rats. Nevertheless, in mammalian brain across species ascorbate is present in high concentrations, up to 0.4 mM in the cerebrospinal fluid and 1–10 mM in glial and neuronal compartments, respectively, due to the presence of carrier-mediated transport (Rice, 2000; Rose and Bode, 1993). Moreover, ascorbic acid is unevenly distributed in different brain regions, is released from neurons by various stimuli, and its extracellular and intracellular concentrations are homeostatically regulated (Bigelow *et al.*, 1984; Boutelle *et al.*, 1989; Pierce and Rebec, 1990, 1994). These observations suggest that ascorbic acid, in addition of being an antioxidant and a cofactor for a number of hydroxylating enzymes (Rose and Bode, 1993), may indeed play a physiological role in modulating intercellular signaling in brain. A time course analysis of the effect of ascorbate (0.2 mM) in brain cortical minces revealed that it enhanced both pregnenolone and DHEA in a time-dependent manner (Figure 1.1). Pregnenolone was already increased (+60%) after 5 minutes and reached a peak (+110%) at 15 minutes, while DHEA, which was not altered after 5 minutes, showed a maximal

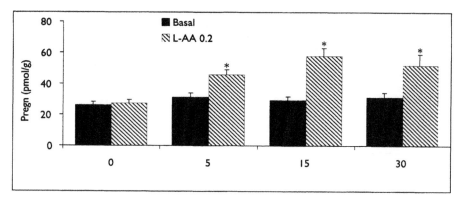

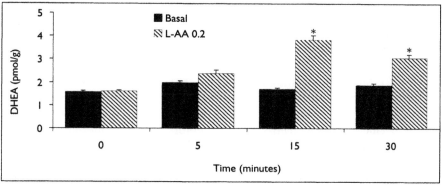

Figure 1.1 Time course of the ascorbic acid effect on pregnenolone and DHEA synthesis/
accumulation in brain cortical minces. Brain cortical minces were incubated for
the indicated time periods with 0.2 mM ascorbic acid (AA). Each bar represents the
mean ± SEM of five individual determinations, each run in duplicate. *$P < 0.05$, vs. the
respective basal value.

increase (+110%) 15 minutes after the exposure of brain tissue to ascorbate. The
ascorbic acid-induced pregnenolone and DHEA increase was concentration-
dependent between 0.07 and 1.0 mM in cerebral cortical minces (Table 1.1). The
incubation of heat-inactivated (95 °C for 10 minutes) brain cortical minces with
0.4 mM ascorbic acid would not affect either pregnenolone or DHEA quantification (Roscetti *et al.*, 1998), suggesting that the effect of ascorbate is not an artefact
due to an increased extraction efficiency, protection of steroids from degradation
during incubation or to an increased reactivity of steroids with their respective
antiserum. In support of the view that the action of ascorbic acid may be of physiological interest, we also found that neurosteroidogenesis was affected by ascorbic
acid in a brain region-dependent manner. Basal neurosteroid concentrations were
higher, with the exception of pregnenolone, in tissue minces from hippocampus
and corpus striatum than cerebral cortex. Pregnenolone was increased by ascorbic
acid in all three brain regions, progesterone and allopregnanolone were increased
only in hippocampus and corpus striatum (Roscetti *et al.*, 1998), and DHEA was
increased in cortex, but failed to be consistently stimulated in hippocampus

Table 1.1 Ascorbic acid concentration dependently stimulates pregnenolone and DHEA synthesis/accumulation in rat brain cortical minces

	Pregnenolone (pmol/g tissue)	DHEA (pmol/g tissue)
Vehicle	28.2±3.5	1.78±0.14
AA 0.07 mM	40.8±4.1	2.1±0.14
AA 0.2 mM	52.8±3.5*	3.2±0.30*
AA 0.4 mM	102±15.2**	6.4±0.7**
AA 1.0 mM	100±9.5**	5.2±0.52**

Notes
Rat brain cortical minces were incubated for 15 minutes with vehicle or various concentrations of ascorbic acid (AA) in a final volume of 5 ml. The tissue minces in each sample were homogenized and the steroids extracted in ethylacetate (1:1, vol:vol). The dried extract, resuspended in N-hexane was purified by HPLC as described (Roscetti et al., 1998). Steroids were quantified by radioimmunoassay with specific antisera. Each value represents the mean ±SEM of ten–fifteen samples, each run in duplicate. *$P<0.05$ and **$P<0.01$, when compared to the respective vehicle value.

(Table 1.2). Although it cannot be excluded that this different sensitivity to ascorbate may depend on brain regional differences in endogenous ascorbate content, the possibility exists, as discussed in the next section, that ascorbate interacts in a region-specific manner with enzymes or receptors that show uneven brain or cellular distribution. The modulation of neurosteroid synthesis/accumulation by ascorbic acid, which also plays a role in the ACTH-dependent adrenal steroidogenesis (Goralczyk *et al.*, 1992), is not affected by the concentration of circulating ACTH present in the rats at the time they are killed. The removal of peripheral steroidogenic tissues – adrenals and gonads – which results in an increase of ACTH release from the pituitary, or a treatment with dexamethasone (1 mg/Kg, 4 days), which inhibits ACTH release from the pituitary, greatly reduced the concentrations of steroids in plasma and, although by a significantly lower extent, also in brain (Roscetti *et al.*,

Table 1.2 Ascorbic acid (AA) stimulates neurosteroid synthesis/accumulation in a brain-region dependent manner

	Pregnenolone (pmol/g tissue)	DHEA (pmol/g tissue)	Progesterone (pmol/g tissue)
Cortex			
Vehicle	33.2±3.9	1.8±0.27	35.3±7.0
AA 1.0 mM	109±16**	5.6±0.6**	33±6.8
Hippocampus			
Vehicle	29±5.7	8.4±1.1	120±11
AA 1.0 mM	116±19**	8.6±1.5	197±18*
Striatum			
Vehicle	68±14	10.5±2.7	127±14
AA 1.0 mM	98±19	20±3.5	208±31*

Notes
Each value represents the mean ±SEM of eight–ten individual determinations, each run in duplicate. *$P<0.05$ and **$P<0.01$, when compared to the respective vehicle values.

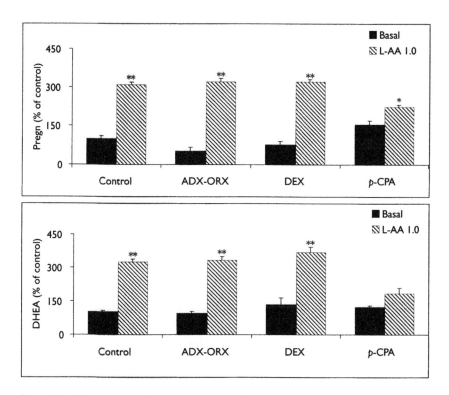

Figure 1.2 Effect of adrenalectomy/orchiectomy (ADX/ORX), dexamethasone (DEX) or p-chlorophenylalanine (p-CPA) pretreatment on the ascorbic acid (AA)-induced pregnenolone and DHEA stimulation in brain cortical minces. The rats were adrenalectomized/orchiectomized 1 week prior to the experiment; dexamethasone-sulfate (1 mg/kg, i.p.) was administered once a day for 4 consecutive days; p-CPA (400 mg/kg, i.p.) was administered for 2 days and the rats were killed 24 hours after the second injection. Each bar represents the mean of four–six individual determinations. **$P < 0.01$ and *$P < 0.05$, when compared to the respective basal value.

1994). However, ascorbic acid, similarly to dibutyryl-cyclic AMP (Roscetti *et al.*, 1994) retained its ability to increase pregnenolone synthesis/accumulation in brain cortical minces of adrenalectomized/orchiectomized (ADX/ORX) or dexamethasone-treated rats (Roscetti *et al.*, 1998). Similarly, DHEA, whose basal brain cortical levels unlike those of pregnenolone were not decreased by either treatment, was enhanced by ascorbate in ADX/ORX and dexamethasone-treated rats to the same extent as in intact, sham-operated or vehicle-treated rats (Figure 1.2). Isoascorbic acid, the D-isomer of ascorbic acid which has similar antioxidant activity and is equally active as a cofactor in hydroxylation reactions as the L-isomer, but is not as good as a substrate for the Na^+-dependent ascorbate transporter, also enhanced pregnenolone and DHEA accumulation in brain cortical slices (Table 1.3). This may suggest that ascorbic acid is acting at the extracellular level; however, it should also be considered that in our *in vitro* experimental conditions there might be a lower competition for the transporter between isoascorbic acid and endogenous ascorbate,

Table 1.3 Isoascorbic acid (Iso-AA) stimulates pregnenolone and DHEA
synthesis/accumulation in rat brain cortical minces

	Pregnenolone (pmol/g tissue)	DHEA (pmol/g tissue)	Progesterone (pmol/g tissue)
Vehicle	32.9±4.7	1.7±0.3	19±2.5
AA 0.2 mM	86±12*	3.5±0.15*	24±2.5
Iso-AA 0.2 mM	80±19*	3.7±0.28*	28±3.6

Note
Each value represents the mean ±SEM of four individual determinations, each run in
duplicate. *$P < 0.05$, when compared to the respective vehicle value.

due to alterations in the ratio between tissue and extracellular ascorbic acid concentrations (Rice, 2000). Ascorbate is also a cofactor for several hydroxylating
enzymes, and, consistent with the fact that the hydroxylation of L-arginine is the
first step in nitric oxide (NO) synthesis, has been shown to stimulate neuronal and
endothelial NO-synthase (Heller *et al.*, 1999; Hofmann and Schmidt, 1995). While
the preincubation of brain cortical minces with N^G-nitro-L-arginine methyl ester
(L-NAME, 100 µM), an inhibitor of NO-synthase, failed to affect both the basal
and ascorbic acid-induced neurosteroid increase, the preincubation with the NO-
donor *S*-nitroso-*N*-acetylpenicillamine (SNAP, 300 µM) inhibited the stimulation
of pregnenolone and DHEA induced by ascorbic acid (Figure 1.3). These results,
in agreement with a previously shown inhibitory action of NO on peripheral
steroidogenesis (Vega *et al.*, 1998; Skarzynski and Okuda, 2000), appear to exclude
that the ascorbic acid-induced neurosteroid stimulation in brain cortical tissue is
mediated by an increase in NO.

The stimulation of neurosteroidogenesis by ascorbic
acid is inhibited by serotonin receptor antagonists

Only serotonin, among the monoamine receptor agonists tested, was able to
increase pregnenolone in brain cortical minces, although with a lower efficacy,
relative to ascorbic acid. A possible explanation for the low efficacy of serotonin
may reside in its interaction with multiple serotonin receptor subtypes expressed
by cerebrocortical cells (Hoyer *et al.*, 1994), that might have opposing effects on
neurosteroidogenesis. Indeed, in adult rat brain, serotonin may stimulate (Barbaccia
et al., 1983) as well as inhibit (De Vivo and Maayani, 1986) the activity of adenylate
cyclase and may, as well, stimulate phospholipase C (Conn and Sanders-Bush, 1985).
Among the serotonin receptor agonists tested, 5-methoxytryptamine (0.4 µM), and
1-(3-chlorophenyl)piperazine (*m*-CPP, 1.0 µM) increased pregnenolone (+60 and
+90%, respectively), but not DHEA, in brain cortical minces. 5-Carboxamido-
tryptamine and the selective $5HT_2$ receptor agonist MK-212 failed to show any
effect. The co-incubation of brain cortical minces with a submaximal concentration
of ascorbic acid and serotonin failed to show an additive action of the two agents
(Roscetti *et al.*, 1998). An interaction between ascorbic acid and serotonergic transmission was suggested by the results of an experiment in which the ability of ascorbic
acid to stimulate pregnenolone and DHEA synthesis/accumulation was tested in

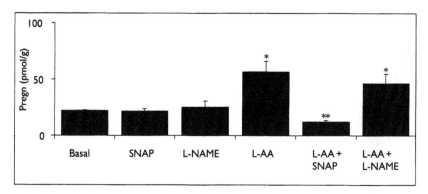

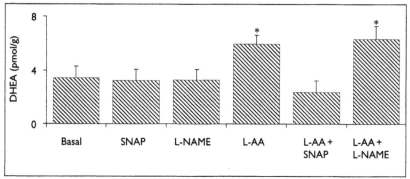

Figure 1.3 Effect of S-nitroso-N-methylpenicillamine (SNAP) and N^G-nitro-L-arginine methyl ester (L-NAME) on the ascorbic acid-induced increase of pregnenolone and DHEA in brain cortical minces. Brain cortical minces were incubated with L-NAME (100 μM) or with SNAP (300 μM) for 15 minutes prior to the addition of ascorbic acid (0.2 mM). Each bar represents the mean ±SEM of six individual determinations. *$P < 0.05$ and **$P < 0.05$, when compared to the respective value in the absence of ascorbic acid (AA).

brain cortical minces of rats that had been depleted of endogenous serotonin by the administration of *p*-chlorophenylalanine, *p*-CPA – 400 mg/kg, i.p. 2 days – an inhibitor of tryptophan-hydroxylase activity which is a rate-limiting step in serotonin synthesis (Koe and Weissman, 1966). The basal pregnenolone and DHEA concentrations were higher, though not reaching statistical significance, and ascorbic acid was significantly less efficacious in increasing the concentrations of both steroids in *p*-CPA-treated rats (Figure 1.2). The effect of ascorbic acid in stimulating neurosteroid synthesis/accumulation was challenged in the presence of several serotonin receptor antagonists: their rank order of potency in inhibiting the effect of ascorbic acid (methiothepin > pizotifen ≥ clozapine > methysergide), the lack of effect of spiperone, MDL 72222 (an antagonist of $5HT_3$ receptors), mesulergine as well as of DL-propranolol, prazosin and S(–)-sulpiride (Figure 1.4) and the relative selectivity of agonists, suggests that ascorbic acid may selectively amplify the action of serotonin at serotonin receptors ($5HT_6$) which are positively coupled to adenylate cyclase (Hoyer *et al.*, 1994).

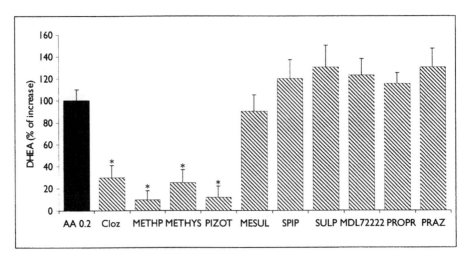

Figure 1.4 Serotonin receptor antagonists inhibit the ascorbic acid-induced DHEA increase in brain cortical minces. Cloz = clozapine (300 nM); METHP = metiothepin (50 nM); METHYS = methysergide (2 μM); PIZOT = pizotifen (1 μM); MESUL = mesulergine (200 nM); SPIP = spiperone (10 μM); SULP = L-sulpiride (1 μM); MDL72222 (1 μM); PROPR = propranolol (1 μM); PRAZ = prazosin (5 μM). Each drug was added 5 minutes before AA = ascorbic acid (0.2 mM). *$P < 0.05$, when compared to the ascorbic acid value.

Inhibition of adenylate cyclase attenuates the effect of ascorbic acid

The overlapping pharmacology, with respect to agonists and antagonists, of the various metabotropic serotonin receptors (Hoyer *et al.*, 1994), and the possibility that some of the antagonists tested would inhibit ascorbic acid-induced neurosteroid formation due to redox properties of the molecule prompted us to further characterize the mechanism(s) involved in the action of ascorbate. Brain cortical minces that had been preincubated with either an inhibitor of adenylate cyclase (MDL 12330, 30 μM) or with an inhibitor of phospholipase C (U-73122, 10 μM) were challenged with ascorbic acid. While U-73122 decreased the basal pregnenolone concentrations, but left unabated the stimulation by ascorbic acid, the inhibition of adenylate cyclase by MDL 12330 failed to affect the basal pregnenolone and DHEA concentrations, but drastically reduced the stimulatory effect of ascorbic acid (Table 1.4). The small increase of pregnenolone accumulation elicited by ascorbic acid in the presence of MDL 12330 may be due to the inhibitory activity of the latter on cyclic AMP phosphodiesterase. At present, we have no clear explanation as to the type of interaction between ascorbic acid and adenylate cyclase-coupled serotonin receptors. It has recently been reported that the incubation of rat or human glial cell lines with 1–30 mM FeSO$_4$, which generates reactive oxygen species, also results in an increased production of DHEA, associated, depending on the cell line, with an increased production of pregnenolone. These

Table 1.4 A preincubation with MDL 12330, an inhibitor of adenylate cyclase, but not with U-73122, an inhibitor of phospholipase C, antagonizes the pregnenolone and DHEA stimulation induced by ascorbic acid

	Pregnenolone (pmol/g tissue)		DHEA (pmol/g tissue)	
	Vehicle	AA 0.4 mM	Vehicle	AA 0.4 mM
Vehicle	24±5.0	94±14*	1.9±0.31	5.6±0.73*
MDL 12330	22±2.4	43±2.9***	2.2±0.28	3.2±0.38**
U-73122	14±1.0	77±7.0*	2.1±0.11	4.5±0.28*

Notes
The brain cortical minces were preincubated for 10 minutes with MDL 12330 (30 μM) or with U-73122 (10 μM), after which vehicle or ascorbic acid was added and the incubation was continued for 15 minutes. Each value represents the mean ± SEM of five individual determinations, each run in duplicate. $*P < 0.01$, when compared to the respective value in the absence of ascorbic acid (AA); $**P < 0.05$, when compared to the vehicle + AA value.

authors suggest that DHEA synthesis in neural cells may occur via a pathway which differs from the one mediated by P450 c17 activity in adrenals and gonads. This alternative pathway should proceed from the free-radical-induced production of (hydroxy)-peroxide precursors of DHEA and/or pregnenolone (Cascio *et al.*, 1998; Brown *et al.*, 2000). *In vivo* the antioxidant and free-radical scavenger properties of ascorbic acid seem predominant (Rose and Bode, 1993; Rice, 2000); *in vitro*, however, ascorbic acid may act as a prooxidant, and in association with transition metals such as ferrous ions (Fe^{++}) may contribute to the generation of free radicals (Lebel and Bondy, 1990). Thus in our *in vitro* experimental conditions ascorbic acid may increase free-radicals production and this, in turn, may stimulate pregnenolone and DHEA synthesis. However, while the prooxidant action of ascorbate has been associated with a lipid peroxidation-induced increase of adenylate cyclase activity in rat brain synaptosomes (Baba *et al.*, 1981), ascorbic acid failed to alter either the basal or the forskolin-stimulated cyclic AMP accumulation in brain cortical minces. Moreover, it is puzzling that, although the ascorbic acid-mediated increase of DHEA is sensitive to inhibition of adenylate cyclase, either dibutyryl-cyclic AMP or forskolin, both of which increase pregnenolone, fail to enhance DHEA (Barbaccia *et al.*, 1992; Roscetti *et al.*, 1994). Also considering the brain-regional specificity of the ascorbic acid action, our interpretation is that the ascorbate effects on pregnenolone and DHEA occur in contiguous but different cell types and that the cells competent to produce DHEA may be enabled to synthesize it when pregnenolone concentrations are sufficiently high and ascorbate is present.

DHEA IN HUMANS

DHEA(S) has been mostly considered as a precursor of androgen and estrogens, and its conversion to these hormones has been implied in the DHEA metabolic, cardiovascular and anti-tumorigenic effects (Ebeling and Koivisto, 1994; Hinson and Raven, 1999). In humans, at variance with rats, DHEA and DHEAS are among the most abundant circulating adrenal steroids. Their plasma levels begin to rise during the sixth–seventh year of age, a time referred to as "adrenarche", peak on early adulthood and gradually decline thereafter (Orentreich *et al.*, 1984; Labrie *et al.*, 1997a). Individual variability in the plasma DHEA/DHEAS levels notwithstanding, longitudinal studies show that individuals with higher values in their early adult life tend to maintain higher levels also in their late life (Thomas *et al.*, 1994). In elderly subjects (75–85 years of age), males and females, DHEA and DHEAS plasma concentrations are approximately 10–20% of those found at their peak. This trend is clearly different from that of cortisol, indicating that other factors in addition to ACTH release might regulate DHEA synthesis and secretion from the adrenals (Adams, 1985; Hinson and Raven, 1999). Since DHEA and DHEAS are reported to exert neuroprotective effects via an antiglucorticoid action in brain (Cardounel *et al.*, 1999; Kaminska *et al.*, 2000) it has been suggested that the progressive age-related unbalance between DHEA(S) and cortisol might have functional significance in the cognitive impairment observed in the elderly population.

DHEA/DHEAS and neuropsychopathology

The novel non-genomic mechanism(s), responsible for the DHEA and DHEAS induced modulation of neurotransmission through ionotropic ligand-gated receptors (Paul and Purdy, 1992; Lambert *et al.*, 1995; Baulieu and Robel, 1998; Rupprecht and Holsboer, 1999), has provided an additional rationale to investigate possible correlations between changes in the endogenous concentrations of these neuroactive (neuro)steroids and human neuropsychopathology that has been associated with alterations of GABAergic and/or glutamatergic transmission. Since blood is much more accessible than brain specimens or other biological fluids, i.e. cerebrospinal fluid, most of these studies have been focused on searching for alterations in circulating DHEA and DHEAS concentrations. A crucial aspect of this type of study concerns the meaning of differences in circulating steroid concentrations with respect to their significance as an index of brain concentrations and, therefore, with respect to their possible relation to alterations in brain function. The pertinence of this question is evident, given that the brain is equipped with the necessary tools to produce neurosteroids and that there might be brain-intrinsic regulatory mechanisms for neurosteroid synthesis/metabolism (see previous sections in this chapter). The latter may be responsible of the fact that, even in non-adrenalectomized animals, brain concentrations of distinct neurosteroids may not always change in concert with plasma levels (Barbaccia *et al.*, 1997; Concas *et al.*, 1998; Barbaccia *et al.*, 1999; Matsumoto *et al.*, 1999). Nevertheless, it should not be neglected that (i) steroids, in particular non-conjugated steroids such as DHEA, easily cross the blood–brain barrier, and, actually, rapidly reach higher concentrations in brain tissue than in plasma (Robel and Baulieu, 1995); (ii) the concentrations of several steroids in human brain tissue specimens, although differing from one brain region to another, appear to correlate with the endocrine status at the moment of death (Bixo *et al.*, 1997); (iii) physiological fluctuations in circulating levels of neuroactive steroids in fertile women appear to impact on brain cortical function. Indeed, decreased cortical excitability has been observed during the luteal phase, i.e. when the circulating levels of progesterone and its $GABA_A$-receptor active metabolite allopregnanolone are higher, compared to the follicular phase (Smith *et al.*, 1999). Thus, although studies in laboratory animals indicate that local synthesis/metabolism influences brain neurosteroid concentrations, changes in circulating neuroactive steroids may also reasonably reflect changes in brain and higher blood concentrations of precursors may contribute to an enhanced synthesis of neurosteroids in brain. This reasoning may not apply to sulfated steroids such as DHEAS, whose polarity greatly reduces its diffusion from blood to brain. In fact, the ratio between the cerebrospinal fluid and serum concentrations of DHEAS is approximately 50-fold lower with respect to that of DHEA (Guazzo *et al.*, 1996).

Neuroactive steroids and menopause-related mood alterations

Differences in circulating, and therefore brain, concentrations of several neuro-active steroids have been hypothesized to have a role in the expression of mood alterations that characterize premenstrual dysphoric disorder and post-partum depression. A subpopulation of women in menopause also report psychological symptoms. These are present mostly during the first three to five years of post-menopause and are represented by negative mood, anxiety, irritability, fatigue, insomnia and forgetfulness (Schmidt *et al.*, 1998). Although it is not clear whether such manifestations are a consequence of the physical distress of this period of life or represent an exacerbation of a latent mood disorder, it has been hypothesized that an imbalance between androgens and estrogens or a reduced psychotropic action of sex steroids and/or their metabolites may play a role in the mood disturbances characteristic of menopause (Gitlin and Pasnau, 1989; Pariser, 1993; Hay *et al.*, 1994; Pearlstein, 1995; Schmidt *et al.*, 1998). This inference is corroborated by the observation that lower cerebrospinal fluid concentrations of allopregnanolone, a potent positive allosteric modulator of the $GABA_A$ receptor endowed with anxiolytic activity, were measured in untreated patients suffering from major depression associated with anxiety with respect to control subjects (Uzunova *et al.*, 1998). These authors also showed that clinical recovery, following a succesful antidepressant treatment, is associated with an enhancement of CSF allopregnanolone concentrations.

While the circulating levels of progesterone and allopregnanolone are dramatically reduced in menopause, due to the cessation of ovarian activity, the circulating levels of DHEA and DHEAS are independent from the steroidogenic activity of the ovary (Jones and James, 1987). Thus, we investigated the possibility that differences not only in the endogenous concentrations of progesterone and allopregnanolone, but also of DHEA, DHEAS and 5α-androstane-3α,17β-diol (3α-ADIOL), a DHEA metabolite that, like allopregnanolone, positively modulates the action of GABA at the $GABA_A$ receptor complex (Gee *et al.*, 1988; Morrow *et al.*, 1990; Frye *et al.*, 1996), may correlate with the presence and intensity of menopause-associated psychological symptoms. The latter were assessed on the basis of the scores obtained according to the Zung Self-rating Depression Scale (ZSDS) and to the Cornell Dysthymia Rating Scale (CDRS) in a group of twenty five post-menopausal (1–3 years) women, which were medication-free and without record of endocrine, metabolic or psychiatric disorders. Thirteen women were classified as "asymptomatic"

Table 1.5 Cornell's Dysthymia Rating Scale (CDRS) scores relative to anxiety-related items (11–15) in asymptomatic and symptomatic menopausal women

	Psychic anxiety	Somatic anxiety	Worries	Irritability	Somatic symptoms
Asymptomatic	1.12±0.2	0.67±0.17	0.55±0.17	0.89±0.2	0.44±0.17
Symptomatic	3±0.3**	2.5±0.38**	1.37±0.46	1.75±0.3*	1.62±0.42*

Notes
Asymptomatic: n = 13; symptomatic: n = 12. Data are expressed as mean ±SEM. *$P<0.02$ and **$P<0.005$, Mann-Whithney-U-Wilcoxon test.

and twelve as "symptomatic" (Barbaccia *et al.*, 2000). A significant ($P < 0.001$) and positive correlation ($r = 0.864$) was observed in the total sample between the ZSDS and CDRS ratings, both indicative of mild depression/dysthymia in the symptomatic group. Moreover, the CDRS, formulated in a two-factor analytic structure (12 items for dysthymic and 5 items for anxiety symptoms), allows to evaluate, on a rating scale from 0 to 4, also anxiety symptoms. Table 1.5 shows that the symptomatic women scored significantly higher than the asymptomatic ones also with respect to anxiety symptoms. The two groups failed to show differences in the plasma levels of prolactin, FSH, LH, cortisol, 17β-estradiol, free and total testosterone, androstenedione and pregnenolone. Also the levels of progesterone and allopregnanolone, which were consistent with the menopausal status were very low, were similar in the two groups (Barbaccia *et al.*, 2000). However, the mean plasma DHEA concentrations were significantly higher in the asymptomatic women (Figure 1.5), while the concentrations of DHEAS, though also showing slightly higher values in the asymptomatic group, failed to show significant differences (Barbaccia *et al.*, 2000). The finding of lower anxiety symptomatology associated with higher

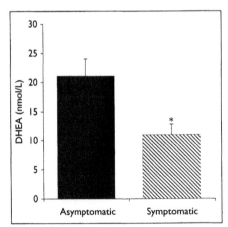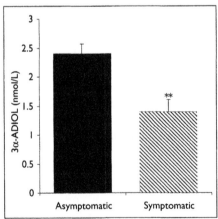

Figure 1.5 DHEA and 3α-ADIOL plasma concentrations in asymptomatic and symptomatic menopausal women. Blood was withdrawn by venipuncture at 8.00 a.m. Asymptomatic: n = 13 and n = 9, symptomatic: n = 12 and n = 8, for the determination of DHEA and 3α-ADIOL, respectively. *$P < 0.05$ and **$P < 0.005$, when compared to the respective asymptomatic value.

Table 1.6 Significant and negative correlation between CDRS anxiety
scores and 3α-ADIOL serum concentrations in the total
population group

	r^a	P value	n
CDRS anxiety/3α-ADIOL	−0.672	<0.005	17
CDRS anxiety/DHEA	−0.412	=0.0503	25
CDRS anxiety/DHEAS	−0.0564	NS	25

Notes
[a] Pearson's correlation coefficient. NS = not significant. Multiple step regression
analysis revealed that neither DHEA nor DHEAS affect the potency or the
significance of the correlation between CDRS/anxiety and 3α-ADIOL.

DHEA plasma levels in the asymptomatic group may appear at odds with an
action of DHEA in brain as positive allosteric modulator of the ionotropic
glutamate receptor of the NMDA type (Paul and Purdy, 1992; Compagnone and
Mellon, 1998) or as negative modulator of $GABA_A$ receptors (Imamura and
Prasad, 1998), mechanisms through which DHEA should, eventually, yield anxiety
(Haefely, 1994; Wiley, 1997). On the other hand, studies in rodents have shown
that the administration of DHEA elicits anxiolytic effects (Melchior and Ritzman,
1994; Prasad et al., 1997) and enhances the hypnotic effect of ethanol and pentobarbital (Melchior and Ritzman, 1992). These actions, not consistent with a clear
cut potentiation of NMDA-receptor function or a negative modulation of $GABA_A$
receptor, could instead be partially explained by the in vivo conversion of DHEA
to 3α-ADIOL and, possibly androsterone, that through a positive modulation of
$GABA_A$ receptors could have an anxiolytic action (Costa et al., 1994; Gee et al.,
1995; Lambert et al., 1995). This inference was verified by measuring 3α-ADIOL,
which was determined only in 17 subjects (nine asymptomatic and eight symptomatic). As shown in Figure 1.5, the 3α-ADIOL serum concentrations, overall in the
same range of those previously reported (Labrie et al., 1997a), were significantly
higher in the asymptomatic group. A positive and highly significant correlation
was observed between the values of DHEA and 3α-ADIOL ($r = 0.814$, $P < 0.001$),
indicating a precursor–product relationship between the two steroids. Moreover,
as shown in Table 1.6, we found a highly significant and negative correlation
between the cumulative scores in the 5 anxiety-related items of the CDRS and the
morning plasma levels of 3α-ADIOL. A recent report shows that a reduction in the
brain concentrations of allopregnanolone results in a decreased potency of muscimol in gating chloride fluxes through $GABA_A$ receptors in cortical neurons and in a
decreased muscimol-induced sleeping time in mice (Pinna et al., 2000). Thus, the
lower 3α-ADIOL concentrations in symptomatic menopausal women, perhaps in
association with the menopause-related relevant reduction in allopregnanolone
levels, may indeed contribute to a reduced positive modulation of central $GABA_A$
receptors and therefore to the expression of anxiety. Of course, it cannot be ruled
out the possibility that DHEA may act through multiple and, perhaps, as yet
undiscovered mechanism(s) of action and it is possible that menopause or
age-related changes in $GABA_A$ receptor structure and function may take part in
the expression of anxiety. The observed intergroup differences in DHEA and

3α-ADIOL suggest that either DHEA plasma concentrations decline more rapidly with age in the symptomatic women or that the latter belong to the "low DHEA(S)" group of the general population (Thomas *et al.*, 1994). Alternatively, since in rodents social isolation decreases brain and plasma concentrations of neuroactive steroids (Matsumoto *et al.*, 1999; Serra *et al.*, 2000), the lower DHEA and 3α-ADIOL concentrations found in symptomatic women may be a consequence of their anxiety symptomatology. These results, although obtained in a limited number of subjects, on one side suggest that differences in the plasma concentrations of 3α-ADIOL, one of the GABA$_A$ receptor active metabolites of DHEA, may indeed reflect a different availability of this neuroactive steroid also in brain, on the other side may lead to surmise that DHEA supplementation may be useful, at least to control the menopause-related anxiety symptomatology. Thus far, however, the analysis of the effects of DHEA administration on various psychological or even physical parameters in the ageing population resulted in controversial findings (Mortola and Yen, 1990; Morales *et al.*, 1994; Berr *et al.*, 1996; Barnhart *et al.*, 1999; Flynn *et al.*, 1999; Baulieu *et al.*, 2000). In particular, a recent study by Barnhart and colleagues (1999) reported the failure of DHEA supplementation to improve health-related-quality-of-life aspects, including anxiety symptomatology, in perimenopausal women. Apart from differences related to methodological issues (i.e. the choice of perimenopausal vs post-menopausal women) between the two studies, it should also be considered that serum 3α-ADIOL concentrations, which at variance with those of DHEA fail to show a significant decrease with age in women (Labrie *et al.*, 1997a), were not increased following daily application of DHEA to the skin for two weeks, a condition that enhances by three-fold DHEA in plasma (Labrie *et al.*, 1997b). These observations suggest that (i) endogenous and exogenously administered DHEA may have a different metabolic fate; (ii) the concentration of endogenous DHEA is not rate limiting for the synthesis of 3α-ADIOL, which may instead be regulated by other factors (for instance, activity or expression of enzymes); (iii) some "*in vivo*" actions of DHEA on the central nervous system may depend on its conversion to neuroactive metabolites.

CONCLUDING REMARKS

The results summarized in this chapter support the view that DHEA is a neurosteroid and its synthesis/accumulation appears to be rapidly regulated by brain-intrinsic mechanisms involving the activation of intracellular second messenger pathways and ascorbic acid. The extracellular concentrations of the latter are regulated by glutamate–ascorbate heteroexchange such that reuptake of glutamate, released by activation of glutamatergic pathways, results in an increase of extracellular ascorbate (Miele *et al.*, 1994; Pierce and Rebec, 1994). Should the ability of ascorbic acid in stimulating DHEA synthesis/accumulation be confirmed also *in vivo*, the interplay between glutamate, ascorbic acid and DHEA, which in turn may interact with ionotropic glutamate receptors of the NMDA-type (Monnet *et al.*, 1995; Maurice *et al.*, 1998; Compagnone and Mellon, 1998), may be relevant in several pathophysiological aspects in which glutamatergic neurotransmission is involved. It is puzzling that, although DHEA appears to exert a number of actions in rodents, i.e. as neuroprotective agent (Kimonides *et al.*, 1998; Cardounel *et al.*, 1999; Kaminska *et al.*, 2000) or as enhancer of neuronal plasticity, both *in vitro* and *in vivo* (Bologa *et al.*, 1987; Flood *et al.*, 1988; Randall *et al.*, 1995; Bergeron *et al.*, 1996; Debonnel *et al.*, 1996; Compagnone and Mellon, 1998; Maurice *et al.*, 1998), thus far, conflicting results have been obtained in humans when trying either to correlate endogenous DHEA(S) with cognitive function or to assess an effect of DHEA administration on cognition or more generally on well-being (Wolf and Kirschbaum, 1999). Consistent with these discrepancies, both groups of women analyzed in our study failed to show cognitive impairment, assessed by Mini Mental State Scale, and, yet, they signifiantly differed for their plasma DHEA levels. However, the lower DHEA and 3α-ADIOL plasma concentrations we have measured in symptomatic menopausal women and mostly the negative correlation found in our group of subjects between intensity of anxiety symptomatology and 3α-ADIOL, whose action as positive modulator of the GABA$_A$ receptor would be consistent with its potential role as an endogenous anxiolytic, suggest that if DHEA has effects in the human central nervous system these may be mediated, at least in some cases, by its neuroactive metabolites, which should not be overlooked when searching for correlations between endogenous DHEA and various aspects of brain function.

REFERENCES

Adams, J.B. (1985) Control of secretion and the function of C_{19}-Δ^5-steroids of the human adrenal gland. *Mol. Cell. Endocrinol.*, **41**, 1–17.

Baba, A., Lee, E., Ohta, A., Tatsuno, T. and Iwata, H. (1981) Activation of adenylate cyclase of rat brain by lipid peroxidation. *J. Biol. Chem.*, **256**, 3679–3684.

Barbaccia, M.L., Brunello, N., Chuang, D.-M. and Costa, E. (1983) Serotonin elicited amplification of adenylate cyclase activity in hippocampal membranes. *J. Neurochem.*, **40**, 1671–1679.

Barbaccia, M.L., Roscetti, G., Trabucchi, M., Ambrosio, C. and Massotti, M. (1992) Cyclic AMP-dependent increase of steroidogenesis in rat brain cortical minces. *Eur. J. Pharmacol.*, **219**, 485–486.

Barbaccia, M.L., Roscetti, G., Trabucchi, M., Purdy, R.H., Mostallino, M.C., Concas, A. *et al.* (1997) The effects of inhibitors of GABAergic transmission and stress on brain and plasma allopregnanolone concentrations. *Brit. J. Pharmacol.*, **120**, 1582–1588.

Barbaccia, M.L., Affricano, D., Trabucchi, M., Purdy, R.H., Colombo, G., Agabio, R. *et al.* (1999) Allopregnanolone and THDOC brain concentrations in Sardinian alcohol-preferring rats. *American Society Neuroscience, Abstract*, **29**, 2049.

Barbaccia, M.L., Lello, S., Sidiropoulou, T., Cocco, T., Sorge, R.P., Cocchiarale, A. *et al.* (2000) Plasma 5α-androstane-3α,17β-diol and endogenous steroid that positively modulates GABA$_A$ receptor function and anxiety: a study in menopausal women. *Psychoneuro-endocrinol.*, **25**, 659–675.

Barnhart, K.T., Freeman, E., Grisso, J.A., Rader, D.J., Sammel, M., Kapoor, S. *et al.* (1999) The effect of dehydroepiandrosterone supplementation to symptomatic perimenopausal women on serum endocrine profiles, lipid parameters, and health-related quality of life. *J. Clin. Endocr. Metab.*, **84**, 3896–3902.

Baulieu, E.-E. and Robel, P. (1998) Dehydroepiandrosterone (DHEA) and dehydro-epiandrosterone sulfate (DHEAS) as neuroactive neurosteroids. *Proc. Natl. Acad. Sci. USA*, **95**, 4089–4091.

Baulieu, E.-E., Thomas, G., Legrain, S., Lahlou, N., Roger, M., Debuire, B. *et al.* (2000) Dehydroepiandrosterone (DHEA), DHEA sulfate, and aging: contribution of the DHEAge study to a sociobiomedical issue. *Proc. Natl. Acad. Sci. USA*, **97**, 4279–4284.

Bergeron, R., de Montigny, C. and Debonnel, G. (1996) Potentiation of neuronal NMDA response induced by dehydroepiandrosterone and its suppression by progesterone: effects mediated via sigma receptors. *J. Neurosci.*, **16**, 1193–1202.

Berr, C., Lafont, S., Debuire, B., Dartigues, J.-F. and Baulieu, E.-E. (1996) Relationship of dehydroepiandrosterone sulfate in the elderly with functional, psychological, and mental status, and short term mortality: a French community-based study. *Proc. Natl. Acad. Sci. USA*, **93**, 13410–13415.

Besman, M.J., Yanagibashi, K., Lee, T.D., Hall, P.F. and Shively, J.E. (1989) Identification of des-(Gly-Ile)-endozepine as an effector of corticotropin-dependent adrenal steroidogenesis: stimulation of cholesterol delivery is mediated by peripheral benzodiazepine receptor. *Proc. Natl. Acad. Sci. USA*, **86**, 4897–4901.

Bigelow, J.C., Brown, D.S. and Wightman, R.M. (1984) γ-Aminobutyric acid stimulates the release of endogenous ascorbic acid from rat striatal tissue. *J. Neurochem.*, **42**, 41–419.

Bixo, M., Andersson, A., Winblad, B., Purdy, R.H. and Backstrom, T. (1997) Progesterone, 5alpha-pregnane-3,20-dione and 3alpha-hydroxy-5alpha-pregnane-20-one in specific regions of the human female brain in different endocrine states. *Brain Res.*, **764**, 173–178.

Bologa, L., Sharma, J. and Roberts, E. (1987) Dehydroepiandrosterone and its sulfate derivative reduce neuronal death and enhance astrocytic differentiation in brain cell cultures. *J. Neurosci. Res.*, **17**, 225–234.

Boutelle, M.G., Svensson, L. and Fillenz, M. (1989) Rapid changes in striatal ascorbate in response to tail-pinch monitored by constant potential voltammetry. *Neuroscience*, **30**, 11–17.

Brown, R.C., Cascio, C. and Papadopoulos, V. (2000) Pathways of neurosteroid biosynthesis in cell lines from human brain: regulation of dehydroepiandrosterone formation by oxidative stress and β-amyloid peptide. *J. Neurochem.*, **74**, 847–859.

Cardounel, A., Regelson, W. and Kalimi, M. (1999) Dehydroepiandrosterone protects hippocampal neurons against neurotoxin-induced cell-death: mechanism of action. *Proc. Soc. Exp. Biol. Med.*, **222**, 145–149.

Cascio, C., Prasad, V.V.K., Lin, Y.Y., Lieberman, S. and Papadopoulos, V. (1998) Detection of P40 c17-independent pathways for dehydroepiandrosterone (DHEA) biosynthesis in brain glial tumor cells. *Proc. Natl. Acad. Sci. USA*, **95**, 2862–2867.

Cavallaro, S., Korneyev, A., Guidotti, A. and Costa, E. (1992) Diazepam-binding inhibitor (DBI)-processing products, acting at the mitochondrial DBI receptor, mediate adreno-corticotropic hormone-induced steroidogenesis in rat adrenal gland. *Proc. Natl. Acad. Sci. USA*, **89**, 10598–10602.

Celotti, F., Melcangi, R. and Martini, L. (1992) The 5α-reductase in the brain: Molecular aspects and relation to brain function. *Front. Neuroendocrinol.*, **13**, 163–215.

Compagnone, N.A., Bulfone, A., Rubenstein, J.L.R. and Mellon, S.H. (1995) Steroidogenic enzyme P450 c17 is expressed in the embryonic central nervous system. *Endocrinology*, **136**, 5212–5223.

Compagnone, N.A. and Mellon, S.H. (1998) Dehydroepiandrosterone: a potential signalling molecule for neocortical organization during development. *Proc. Natl. Acad. Sci. USA*, **95**, 4678–4683.

Concas, A., Mostallino, M.C., Porcu, P., Follesa, P., Barbaccia, M.L., Trabucchi, M. *et al.* (1998) Role of brain allopregnanolone in the plasticity of γ-aminobutyric acid type A receptor in rat brain during pregnancy and after delivery. *Proc. Natl. Acad. Sci. USA*, **95**, 13284–13289.

Conn, P.J. and Sanders-Bush, E. (1985) Serotonin-stimulated phosphoinoditide turnover: mediation by the S_2 binding site in rat cerebral cortex but not in subcortical regions. *J. Pharmacol. Exp. Ther.*, **234**, 195–203.

Corpechot, C., Robel, P., Axelson, M., Sjovall, J. and Baulieu, E.-E. (1981) Characterization and measurement of dehydroepiandrosterone sulfate in rat brain. *Proc. Natl. Acad. Sci. USA*, **78**, 4704–4707.

Costa, E., Cheney, D.L., Grayson, D.R., Korneyev, A., Longone, P., Pani, L. *et al.* (1994) Pharmacology of neurosteroid biosynthesis: role of the mitochondrial DBI receptor (MDR) complex. In E.R. de Kloet, A.C. Azmitia and P.W. Landfield (eds), *Brain Corticosteroid Receptors. Ann. N. Y. Acad. Sci.*, **746**, 223–242.

Debonnel, G., Bergeron, R. and de Montigny, C. (1996) Potentiation by dehydroepiandrosterone of the neuronal response to *N*-methyl-D-aspartate in the CA3 region of the rat dorsal hippocampus: an effect mediated via sigma receptors. *J. Endocrinol.*, **150**, S33-S42.

De Vivo, M. and Maayani, S. (1986) Characterization of 5-hydroxytryptamine$_{1a}$ receptor-mediated inhibition of forskolin-stimulated adenylate cyclase activity in guinea pig and rat hippocampal membranes. *J. Pharmacol. Exp. Ther.*, **238**, 248–252.

El-Etr, M., Akwa, Y., Robel, P. and Baulieu, E.-E. (1998) Opposing effects of different steroid sulfates on GABA$_A$ receptor-mediated chloride fluxes. *Brain Res.*, **790**, 334–338.

Ebeling, P. and Koivisto, V.A. (1994) Physiological importance of dehydroepiandrosterone. *Lancet*, **343**, 1479–1481.

Evans, R.M. (1988) The steroid and thyroid hormone receptor superfamily. *Science*, **240**, 889–895.

Flood, J.F., Smith, G.E. and Roberts, E. (1988) Dehydroepiandrosterone and its sulfate derivative enhance memory retention in mice. *Brain Res.*, **447**, 269–278.

Flynn, M.A., Weaver-Osterholtz, D., Sharpe-Timms, K.L., Allen, S. and Krause, G. (1999) Dehydroepiandrosterone replacement in aging humans. *J. Clin. Endocr. Metab.*, **84**, 1527–1533.

Frye, C.A., Duncan, J.E., Basham, M. and Erskine, M.S. (1996) Behavioral effects of 3-alpha androstanediol. II: Hypothalamic and preoptic area actions via a GABAergic mechanism. *Behav. Brain Res.*, **79**, 119–130.

Gee, K.W., Bolger, M.B., Brinton, R.E., Coirini, H. and Mc Ewen, B.S. (1988) Steroid modulation of the chloride ionophore in rat brain: stucture–activity requirements, regional dependence and mechanism of action. *J. Pharmacol. Exp. Ther.*, **246**, 803–812.

Gee, K.W., Mc Cawley, L.D. and Lan, N.C. (1995) A putative receptor for neurosteroids on the GABA$_A$ receptor complex: the pharmacological properties and therapeutic potential of epalons. *Crit. Rev. Neurobiol.*, **9**, 207–227.

Gitlin, M.J. and Pasnau, R.O. (1989) Psychiatric syndromes linked to reproductive function in women: A review of current knowledge. *Am. J. Psychiat.*, **146**, 1413–1422.

Goralczyk, R., Moser, U.K., Matter, V. and Wieser, H. (1992) Regulation of steroid hormone metabolism requires L-ascorbic acid. *Ann. N. Y. Acad. Sci.*, **669**, 349–351.

Guarneri, P., Papadopoulos, V., Pan, B. and Costa, E. (1992) Regulation of pregnenolone synthesis in C6 2B glioma cells by 4′-chlorodiazepam. *Proc. Natl. Acad. Sci. USA*, **89**, 5118–5222.

Guazzo, E.P., Kirkpatrick, P.J., Goodyer, I.M., Shiers, H.M. and Herbert, J. (1996) Cortisol, dehydroepiandrosterone (DHEA) and DHEA sulfate in the cerebrospinal fluid of man: relation to blood levels and the effects of age. *J. Clin. Endocr. Metab.*, **81**, 3951–3960.

Haefely, W.E. (1994) Allosteric modulation of GABA$_A$ receptor channel: a mechanism for interaction with a multitude of central nervous system functions. In H.C. Mohler and M. da Prada (eds), *The Challenge of Neuropharmacology*, pp. 15–39. Éditions Roche, Basel.

Hay, A.G., Bancroft, J. and Johnstone, E.C. (1994) Affective symptoms in women attending a menopause clinic. *Brit. J. Psychiat.*, **164**, 513–516.

Heller, R., Munscher-Paulig, F., Grabner, R. and Till, U. (1999) L-ascorbic acid potentiates nitric oxide synthesis in endothelial cells. *J. Biol. Chem.*, **274**, 8254–8260.

Hinson, J.P. and Raven, P.W. (1999) DHEA deficiency syndrome: a new term for old age? *J. Endocrinol.*, **163**, 1–5.

Hoffman, H. and Schmidt, H.H. (1995) Thiol dependence of nitric oxide synthase. *Biochemistry*, **34**, 13443–13452.

Hoyer, D., Clarke, D.E., Fozard, J.R., Hartig, P.R., Martin, G.R., Mylecharane, E.J. *et al.* (1994) VII. International Union of Pharmacology classification of receptors for 5-hydroxytryptamine (serotonin). *Pharmacol. Rev.*, **46**, 157–203.

Hu, Z.Y., Bourreau, E., Jung-Testas, I., Robel, P. and Baulieu, E.-E. (1987) Neurosteroids: oligodendrocyte mitochondria convert cholesterol to pregnenolone. *Proc. Natl. Acad. Sci. USA*, **84**, 8215–8219.

Imamura, M. and Prasad, C. (1998) Modulation of GABA-gated chloride ion influx in the brain by dehydroepiandrosterone and its metabolites. *Biochem. Bioph. Res. Comm.*, **243**, 771–775.

Jones, D.L. and James, V.H. (1987) Determination of dehydroepiandrosterone and dehydroepiandrosterone sulfate in blood and tissue. Studies in normal women and women with breast or endometrial cancer. *J. Steroid Biochem.*, **26**, 151–159.

Jung-Testas, I., Hu, Z.Y., Baulieu, E.-E. and Robel, P. (1989) Neurosteroids: biosyntesis of pregnenolone and progesterone in primary cultures of rat glial cells. *Endocrinology*, **125**, 2083–2091.

Kaminska, M., Harris, J., Gijsbers, K. and Dubrovsky, B. (2000) Dehydroepiandrosterone sulfate (DHEAS) counteracts decremental effects of corticosterone on dentate gyrus LTP. Implications for depression. *Brain Res. Bull.*, **52**, 229–234.

Karavolas, H.J. and Hodges, D.R. (1991) Metabolism of progesterone and related steroids by neural and neuroendocrine structures. In E. Costa and S.M. Paul (eds), *Neurosteroids and brain function*, Fidia Research Foundation Symposium Series, Thieme, New York, pp. 135–145.

Kimonides, V.G., Khatibi, N.H., Svendsen, C.N., Sofroniew, M.V. and Herbert, J. (1998) Dehydroepiandrosterone (DHEA) and DHEA-sulfate (DHEAS) protect hippocampal neurons against excitatory amino acid-induced neurotoxicity. *Proc. Natl. Acad. Sci. USA*, **95**, 1852–1857.

Koe, B.K. and Weissman, A. (1966) *p*-Chlorophenylalanine: a specific depletor of brain serotonin. *J. Pharmacol. Exp. Ther.*, **154**, 499–516.

Koenig, H.L., Schumacher, M., Farraz, B., Thi, A.N., Ressouches, A., Guennoun, R. *et al.* (1995) Progesterone synthesis and myelin formation by Schwann cells. *Science*, **268**, 1500–1503.

Labrie, F., Belanger, A., Cusan, L., Gomez, J.L. and Candas, B. (1997a) Marked decline in serum concentrations of adrenal C19 sex steroid precursors and conjugated androgen metabolites during aging. *J. Clin. Endocr. Metab.*, **82**, 2396–2402.

Labrie, F., Belanger, A., Cusan, L. and Candas, B. (1997b) Physiological changes in dehydroepiandrosterone are not reflected by serum levels of active androgens and estrogens but of their metabolites: intracrinology. *J. Clin. Endocr. Metab.*, **82**, 2403–2409.

Lambert, J.J., Belelli, D., Hill-Venning, C. and Peters, J.A. (1995) Neurosteroids and GABA$_A$ receptor function. *Trends Pharmacol. Sci.*, **16**, 295–303.

Lebel, C.P. and Bondy, S.C. (1990) Sensitive and rapid quantitation of oxygen reactive species formation in rat synaptosomes. *Neurochem. Int.*, **17**, 435–440.

Le Goascogne, C., Robel, P., Gouezou, P., Sananes, N. and Baulieu, E.-E. (1987) Neurosteroids: cytochrome P-450 scc in rat brain. *Science*, **237**, 1212–1214.

Le Goascogne, C., Sananes, N., Gouezou, M., Takemori, S., Kominami, S., Baulieu, E.-E. *et al.* (1991) Immunoreactive cytochrome P450 c17 in rat and guinea pig gonads, adrenal glands and brain. *J. Reprod. Fertil.*, **93**, 609–622.

Lieberman, S. (1995) An abbreviated account of some aspects of the biochemistry of DHEA, 1934–1995. In F.L. Bellino, R.A. Daynes, P.J. Hornsby, D.H. Lavrin and J.E. Nestler (eds), *Dehydroepiandrosterone (DHEA) and aging*, Ann. N. Y. Acad. Sci., **774**, 1–15, The New York Academy of Sciences, New York.

Majewska, M.D. (1995) Neuronal Actions of Dehydroepiandrosterone. Possible roles in brain development, aging, memory and affect. In F.L. Bellino, R.A. Daynes, P.J. Hornsby, D.H. Lavrin and J.E. Nestler (eds), *Dehydroepiandrosterone (DHEA) and aging, Ann. N. Y. Acad. Sci.*, **774**, 111–120, The New York Academy of Sciences, New York.

Majewska, M.D., Harrison, N.L., Schwartz, R.D., Barker, J.J. and Paul, S.M. (1986) Steroid hormone metabolites are barbiturate-like modulators of the GABA receptor. *Science*, **232**, 1004–1007.

Matsumoto, K., Uzunova, V., Pinna, G., Taki, K., Uzunov, D.P., Watanabe, H. *et al.* (1999) Permissive role of brain allopregnanolone content in the regulation of pentobarbital-induced righting reflex. loss. *Neuropharmacology*, **38**, 955–963.

Maurice, T., Su, T.P. and Privat, A. (1998) Sigma 1 receptor agonists and neurosteroids attenuate β25–35 amyloid peptide-induced amnesia in mice through a common mechanism. *Neuroscience*, **83**, 413–428.

McEwen, B.S. (1991) Non genomic and genomic effects of steroids on neural activity. *Trends Pharmacol. Sci.*, **12**, 141–147.

Melchior, C.L. and Ritzman, R.F. (1992) Dehydroepiandrosterone enhances the hypnotic and hypothermic effects of ethanol and pentobarbital. *Pharmacol. Biochem. Behav.*, **43**, 223–227.

Melchior, C.L. and Ritzman, R.F. (1994) Dehydroepiandrosterone is an anxiolitic in mice on the plus maze. *Pharmacol. Biochem. Behav.*, **47**, 437–441.

Mellon, S.H. and Deschepper, C.F. (1993) Neurosteroid biosynthesis: genes for adrenal steroidogenic enzymes are expressed in the brain. *Brain Res.*, **629**, 283–292.

Mensah-Nyagan, A.G., Do-Rego, J.L., Beaujean, D., Pelletier, G. and Vaudry, H. (1999) Neurosteroids: expression of steroidogenic enzymes and regulation of steroid biosynthesis in the central nervous system. *Pharmacol. Rev.*, **51**, 63–81.

Meyer, J.H., Lee, S., Wittemberg, G.F., Randall, R.D. and Gruol, D.L. (1999) Neurosteroid regulation of inhibitory synaptic transmission in the rat hippocampus *in vitro*. *Neuroscience*, **90**, 1177–1183.

Miele, M., Boutelle, M.G. and Fillenz, M. (1994) The physiologically induced release of ascorbate in rat brain is dependent on impulse traffic, calcium influx and glutamate uptake. *Neuroscience*, **62**, 87–91.

Monnet, F.P., Mahe, V., Robel, P. and Baulieu. E.-E. (1995) Neurosteroids, via sigma receptors, modulate the [^{3}H]norepinephrine release evoked by N-methyl-D-aspartate in the rat hippocampus. *Proc. Natl. Acad. Sci. USA*, **92**, 3774–3778.

Morales, A., Nolan, J., Nelson, J. and Yen, S. (1994) Effects of replacement dose of dehydroepiandrosterone in men and women of advancing age. *J. Clin. Endocr. Metab.*, **78**, 1360–1367.

Morrow, A.L., Pace, J.R., Purdy, R.H. and Paul, S.M. (1990) Characterization of steroid interaction with γ-aminobutyric acid receptor-gated chloride channels: evidence for multiple steroid recognition sites. *Mol. Pharmacol.* **37**, 263–270.

Mortola, J.F. and Yen, S.S.C. (1990) The effects of oral dehydroepiandrosterone on endocrine metabolic parameters in postmenopausal women. *J. Clin. Endocr. Metab.*, **71**, 696–704.

Mukhin, A.G., Papadopoulos, V., Costa, E. and Krueger, K.E. (1989) Motochondrial benzodiazepine receptors regulate steroid biosynthesis. *Proc. Natl. Acad. Sci. USA*, **86**, 9813–9816.

Orentreich, N., Brind, J.L., Ritzler, R.L. and Vogelman, J.H. (1984) Age changes and sex differences in serum dehydroepiandrosterone sulfate concentrations troughout adulthood. *J. Clin. Endocr. Metab.*, **59**, 551–555.

Papadopoulos, V. (1993) Peripheral benzodiazepine/diazepam binding inhibitor receptor: biological role in steroidogenic cell function. *Endocr. Rev.*, **14**, 222–240.

Pariser, S.F. (1993) Women and mood disorders. Menarche to menopause. *Ann. Clin. Psychiat.*, **5**, 249–254.

Park-Chung, M., Malayev, A., Purdy, R.H., Gibbs, T.T. and Farb, D.H. (1999) Sulfated and unsulfated steroids modulate γ-aminobutyric acid A receptor function through distinct sites. *Brain Res.*, **830**, 72–87.

Paul, S.M. and Purdy, R.H. (1992) Neuroactive steroids. *FASEB J.*, **6**, 2311–2322.

Pearlstein, T.B. (1995) Hormones and depression: What are the facts about premenstrual syndrome, menopause and hormone replacement therapy. *Am. J. Obstet. Gynecol.*, **175**, 646–653.

Pierce, R.C. and Rebec, G.V. (1990) Stimulation of both D_1 and D_2 dopamine receptors increases behavioral activation and ascorbate release in the neostriatum of freely moving rats. *Eur. J. Pharmacol.*, **191**, 295–302.

Pierce, R.C. and Rebec, G.V. (1994) A vitamin as neuromodulator: ascorbate release into the extracellular fluid of the brain regulates dopaminergic and glutamatergic transmission. *Prog. Neurobiol.*, **43**, 537–565.

Pinna, G., Uzunova, V., Matsumoto, K., Puia, G., Mienville, J.-M., Costa, E. *et al.* (2000) Brain allopregnanolone regulates the potency of the GABA$_A$ receptor agonist muscimol. *Neuropharmacology*, **39**, 440–448.

Prasad, A., Imamura, M. and Prasad, C. (1997) Dehydroepiandrosterone decreases behavioral despair in high but not low anxiety rats. *Pharmacol. Biochem. Behav.*, **62**, 1053–1057.

Puia, G., Santi, M.-R., Vicini, S., Pritchett, D.B., Purdy, R.H., Paul, S.M. *et al.* (1990) Neurosteroids act on recombinant human GABA$_A$ receptors. *Neuron*, **4**, 759–765

Randall, R.E., Lee, S.Y., Meyer, J.H., Wittenberg, G.F. and Gruol, D.L. (1995) Acute alcohol blocks neurosteroid modulation of synaptic transmission and long-term potentiation in the rat hippocampal slice. *Brain Res.*, **701**, 238–248.

Rice, M.E. (2000) Ascorbate regulation and its neuroprotective role in the brain. *Trends Neurosci.*, **23**, 209–216.

Robel, P. and Baulieu, E.-E. (1995) Dehydroepiandrosterone (DHEA) is a neuroactive neurosteroid. In F.L. Bellino, R.A. Daynes, P.J. Hornsby, D.H. Lavrin and J.E. Nestler (eds), *Dehydroepiandrosterone (DHEA) and aging*, *Ann. N. Y. Acad. Sci.*, **774**, 82–110, The New York Academy of Sciences, New York.

Robel, P., Jung-Testas, I., Hu, Z.Y., Akwa, Y., Sananes, N., Kabbadij, K. *et al.* (1991) Neurosteroids: biosynthesis and metabolism in cultured rodent glia and neuronal cells. In E. Costa and S.M. Paul (eds), *Neurosteroids and brain function*, Fidia Research Foundation Symposium Series, Thieme, New York, pp. 147–154.

Roscetti, G., Ambrosio, C., Trabucchi, M., Massotti, M. and Barbaccia, M.L. (1994) Modulatory mechanisms of cyclic AMP-stimulated steroid content in rat brain cortex. *Eur. J. Pharmacol.*, **269**, 17–24.

Roscetti, G., Del Carmine, R., Trabucchi, M., Massotti, M., Purdy, R.H. and Barbaccia, M.L. (1998) Modulation of neurosteroid synthesis/accumulation by L-ascorbic acid in rat brain tissue: inhibition by selected serotonin antagonists. *J. Neurochem.*, **71**, 1108–1117.

Rose, R.C. and Bode, A.M. (1993) Biology of free radical scavengers: an evaluation of ascorbate. *FASEB J.*, **7**, 1135–1142.

Rupprecht, R. and Holsboer, F. (1999) Neuroactive steroids: mechanisms of action and neuropsychopharmacological perspectives. *Trends Neurosci.*, **22**, 410–416.

Schmidt, P.J., Roca, C.A. and Rubinow, D.R. (1998) Clinical evaluation in studies of perimenopausal women: position paper. *Psychopharmacol. Bull.*, **34**, 309–311.

Serra, M., Madau, P., Chessa, M.F., Caddeo, M., Sanna, E., Trapani, G. *et al.* (1999) 2-Pheny-imidazo-[1,2-a]pyridine derivatives as ligands for peripheral benzodiazepine receptors: stimulation of neurosteroid synthesis and anticonflict action in rats. *Brit. J. Pharmacol.*, **127**, 177–187.

Serra, M., Pisu, M.G., Littera, M., Papi, G., Sanna, E., Tuveri, F. *et al.* (2000) Social isolation-induced decrease in both the abundance of neuroactive steroids and GABA$_A$ receptor function in rat brain. *J. Neurochem.*, **75**, 732–740.

Skarzynski, D.J. and Okuda, K. (2000) Different actions of noradrenaline and nitric oxide on the output of prostaglandins and progesterone in culturad bovine luteal cells. *Prostag. Lipid Mediat.*, **60**, 35–47.

Smith, M.J., Keel, J.C., Greenberg, B.D., Adams, L.F., Schmidt, P.J., Rubinow, D.A. and Wassermann, E.M. (1999) Menstrual cycle effects on cortical excitability. *Neurology*, **53**, 2069–2072.

Sousa, A. and Ticku, M.K. (1997) Interaction of the neurosteroid dehydroepiandrosterone sulfate with the GABA$_A$ receptor complex reveals that it may act via the picrotoxin site. *J. Pharmacol. Exp. Ther.*, **282**, 827–833.

Stromstedt, M. and Waterman, M.R. (1995) Messenger RNAs encoding steroidogenic enzymes are expressed in rodent brain. *Mol. Brain Res.*, **34**, 670–697.

Thomas, G., Frenoy, N., Legrain, S., SebagLanoe, R., Baulieu, E.-E. and Debuire, B. (1994) Serum dehydroepiandrosterone sulfate levels as an individual marker. *J. Clin. Endocr. Metab.*, **79**, 1273–1276.

Uzunova, V., Sheline, Y., Davis, J.M., Rasmusson, A., Uzunov, D.P., Costa, E. *et al.* (1998) Increase in the cerebrospinal fluid content of neurosteroids in patients with unipolar major depression who are receiving fluoxetine or fluvoxamine. *Proc. Natl. Acad. Sci. USA*, **95**, 3239–3244.

Vega, M., Johnson, M.C., Diaz, H.A., Urrutia, L.R., Troncoso, J.L. and Devoto, L. (1998) Regulation of human luteal steroidogenesis *in vitro* by nitric oxide. *Endocrine*, **8**, 185–191.

Wiley, J.L. (1997) Behavioral pharmacology of N-methyl-D-aspartate antagonists: implications for the study and pharmacotherapy of anxiety and schizophrenia. *Exp. Clin. Psychopharmacol.*, **5**, 365–374.

Wolf, O.T. and Kirschbaum, C. (1999) Actions of dehydroepiandrosterone and its sulfate in the central nervous system: effects on cognition and emotion in animals and humans. *Brain Res. Rev.*, **30**, 264–288.

Zwayn, I.H. and Yen, S.S.C. (1999) Dehydroepiandrosterone: biosynthesis and metabolism in the brain. *J. Clin. Endocr. Metab.*, **140**, 880–887.

Chapter 2

DHEA metabolites during the life span

Richard Hampl, Martin Hill and Luboslav Stárka

INTRODUCTION

Dehydroepiandrosterone and its sulfate (DHEA/S) has been attracting attention of scientists for more than four decades due to a number of features by which it differs from other naturally occurring steroids present in humans: as a sulfate it is the most abundant circulating steroid, the concentrations of which are of one or two orders of magnitude higher than those of cortisol. Though it is a typical adrenal steroid, its diurnal rhythm and regulation is not paralleled by that of cortisol. There is no feedback action of DHEA/S on the pituitary or the hypothalamus (Wolf and Kirschbaum, 1999). In contrast to most hormonal steroids the major portion of DHEA is secreted by adrenals in a form of sulfate. In spite of numerous biological effects displayed by DHEA/S at various levels, neither specific receptors for this steroid, nor its competition for other hormonal receptors have been found so far. Besides adrenals and gonads, DHEA/S is synthesized by various neural cells including brain. Brain structures are also its targets. The actions of DHEA/S as neurosteroids are discussed in other chapters in this volume. The typical characteristics of DHEA/S, however, is its decline with age in both sexes, hand in hand with degenerative changes typical for ageing, as muscular weakness, hypertension, osteoporosis, diabetes, susceptibility to infection with an over-all impairment of immune function. The beneficial effects ascribed to DHEA, many of which can be explained by its immunoprotective or antiglucocorticoid effects (Kalimi *et al.*, 1994), rendered this steroid a hot candidate for replacement therapy of seniors. A complex study on 280 subjects above 60 years to whom DHEA was given perorally for one year has been reported recently by Baulieu's group (Baulieu *et al.*, 2000).

Generally, the changes of DHEA/S during life and under various physiological and pathological states have been addressed by many authors. Here we would like to mention only briefly our recent results on DHEA/S and then to focus on several metabolites of DHEA, the importance of which has been revealed recently.

DHEA/S DURING LIFE AND THE IMPORTANCE OF STEROID SULFATASE

Many studies have been devoted to the physiological changes of circulating DHEA/S levels during the human life. We have been following DHEA/S serum concentrations in a representative number of healthy population of both sexes from 1 month until 100 years of age, taking advantage of large population groups examined during the screening of iodine deficiency in our country. The curves reflecting the statistical mean predictions with 95% confidence limits of unconjugated DHEA for women (A) and men (B) are shown on Figure 2.1; the data were obtained by courtesy of Dr J. Šulcová, for further information see Šulcová et al. (1997). In women, the course of the curves of DHEA differed from its sulfate (Šulcová et al., 1997), where additional local maxima could be seen around the age 45 and 60 years of age (Figure 2.1). This fact may be of importance with respect to the different role of unconjugated and sulfated steroids in brain as allosteric regulators of $GABA_A$ receptors (Rupprecht and Holsboer, 1999; Wolf and Kirschbaum, 1999). It points also to the regulatory role of steroid sulfatase for maintaining the level of unconjugated DHEA, which in humans represents less than 1% of the concentration of sulfated steroid and, in addition, its biological half time is 14–20 times shorter than that of DHEAS (30 minutes vs 7–10 hours) (Zumoff et al., 1980). Most of the circulating DHEA is a product of the action of steroid sulfatase action, while only a small amount of DHEA (5–7%) is re-converted by hydroxysteroid sulfotransferase in liver and kidney to DHEAS (Longcope, 1996). Only few reports dealt with the changes of sulfatase activity during life (Cuevas-Covarrubias et al., 1993; Miranda-Duarte et al., 1999). Since the gene of human steroid sulfatase has been cloned (Ballabio and Shapiro, 1995), its expression has been demonstrated in a variety of tissues, where the biological activity of DHEA and its metabolites regulate cell functions (Compagnone et al., 1997). With respect to antiglucocorticoid actions of unconjugated DHEA (Kalimi et al., 1995), inhibiting steroid sulfatase may lead to an anti-inflamatory effect (Suitters et al., 1999). Undoubtedly, steroid sulfatase is an important regulatory factor influencing the availability of biologically active DHEA and of its metabolism, and further studies are needed for investigating the activity of this enzyme during ageing.

DHEA METABOLISM

Concerning the already mentioned beneficial effects of DHEA, attention has recently been focused to DHEA/S itself. This is not surprising when considering its large concentrations in circulation. DHEA and its sulfate, however, undergo an intensive metabolism in most tissues and organs (Baulieu, 1996). One reaction leads to free DHEA through hydrolysis of DHEAS by steroid sulfatase, and the opposite reaction mediated by sulfotransferase. The importance of the former enzyme has been emphasized in the previous paragraph, and deficiency in steroid sulfatase causes ichtyosis, a well known dermal disease (Milone et al., 1991). Steroid sulfatase in concert with sulfotransferase plays an important role especially in the brain tissues, with respect to a limited ability of DHEAS to pass across the blood–brain barrier and the fact that unconjugated DHEA, as well as its precursor pregnenolone act on GABA receptors in a way opposite to that of sulfated steroids.

Besides of the metabolic pathways leading to androstenedione and further to androgens and estrogens, of its oxidoreduction on C_3 and C_{17} leading to isomeric androstenediols or, after reduction of the double bound at C_5, to androstanediols

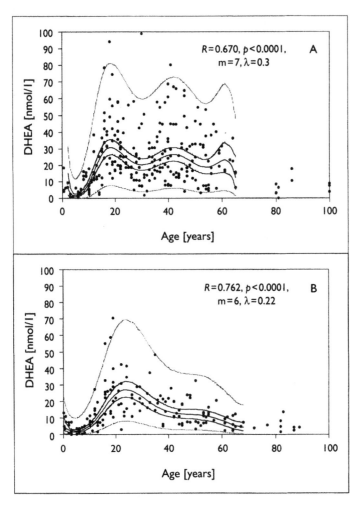

Figure 2.1 Dependence of DHEA levels on age in healthy subjects. Polynomial regression of the multiple degree was employed for the curve fitting. Due to non-Gaussian distribution of the data, the original values were transformed to minimum skewness studentized residues and then re-transformed to the original scale by power transformation. The optimal model was determined from the minimum of Akaike's information criterion. The experimental points with studentized residua greater than 3 were excluded from the computations of mean predictions and the confidence limits as well as the influential points with leverage greater than treble of the average. The bold central curve represents the mean prediction, the thin lines closer to the central one the 95% confidence interval and the more remote curves the 95% confidence interval of prediction after re-transformation of the results to original scale. R denotes the correlation coefficient of the multiple regression, p is the level of statistical significance of the model, m is the degree of polynomial and λ is the power of transformation. A (upper panel): females, B (lower panel): males. Reproduced with kind permission of Dr J. Šulcová from Šulcová et al. (1997).

and androstanedione, a large portion of DHEA is hydroxylated at C_7 and C_{16} into 7α-, 7β-, 16α- and 16β-hydroxyderivatives. In this chapter, we will focus mainly on these latter hydroxylated metabolites, with studies of their changes in ageing.

7-OXYGENATED DHEA METABOLITES

Our group was among the first who discovered 7-hydroxylated metabolites of DHEA in various body fluids (Stárka *et al.*, 1962). Later on we studied the enzymatic 7-hydroxylation in various tissues including fetal ones (Šulcová *et al.*, 1968), though nobody knew at that time, what they were good for. For the history of 7-hydroxy DHEA see our recent minireview (Hampl *et al.*, 1997). The enzyme responsible for 7α-hydroxylation in rat liver was at first described by Šulcová and Stárka (1968). More recent studies including cloning of the gene encoding for the enzyme(s) responsible for 7-hydroxylation activity revealed that 7α- and 7β-hydroxylase are different species, not identical with 7-hydroxylase of cholesterol (for literature see Hampl *et al.*, 1997). The physiological role of 7-OH-DHEA has not yet been entirely elucidated, until recently. In connection with the discovery of non-genomic immunomodulary and antiglucocorticoid effects of DHEA (Kalimi *et al.*, 1994) and the fact that a relatively large dose of this steroid is required to demonstrate its immunoprotective effects, a question arose whether not just DHEA itself, but some of its metabolites were responsible for these actions (Morfin and Courchay, 1994; Padgett and Loria, 1994). Evidence was collected based on *in vitro* experiments with various cell cultures as well as *in vivo* experiments with mice that both 7-hydroxyisomers of DHEA and also their 17β-hydroxymetabolites were able to counteract some immunosuppressive effects of glucocorticoids including dexamethasone-induced apoptosis (Hampl *et al.*, 1997; Chmielewski *et al.*, 2000 and the literature therein). Though evidence was brought that 7α-hydroxysteroids do not compete with glucocorticoids for their receptors (Stárka *et al.*, 1998), recent experiment showed that these metabolites formed in human tonsils, may serve as signals for the T-lymphocyte-induced differentiation of B-cells into specific IgG producing plasma cells (Lafaye *et al.*, 1999). In contrast to tonsillar tissue, T-cells do not 7-hydroxylate DHEA and a paracrine action of these steroids was suggested. In another experiment with human adipose cells, the addition of dexamethasone led to induction of 7α-hydroxylating enzymes (Khalil *et al.*, 1994). This may imply that when circulating glucocorticoids rise, target cells react by increasing the synthesis of 7-hydroxysteroids. An increased 7-hydroxylation of DHEA and also of its precursor pregnenolone at the expense of their oxidation to 4-ene-3-oxosteroids has also been reported in cultured rat astrocytes when the cell contact was increased as during inflammation (Akwa *et al.*, 1993). All of these results indicate that 7-hydroxylation may represent a response to the local impulse and support the idea of their action as locally active agents, counteracting the exaggerated effect of glucocorticoids. As demonstrated in mice, 7α-OH-DHEA is also produced in considerable amounts in brain (Morfin and Courchay, 1994; Doostzadeh and Morfin, 1996). This 7α-hydroxylation was due to expression of the gene encoding for a cytochrome P450 (Cyp7b) (Stapleton *et al.*, 1995; Rose *et al.*, 1997). These data indicate that 7-hydroxylated 5-ene-DHEA- and pregnenolone metabolites may also function as neuroactive steroids. From the point of view of a plausible immunomodulating role of 7-OH-DHEA our most recent detection of both 7-OH-DHEA isomers in human semen, where their concentration exceeds those in blood may be of interest (Hampl *et al.*, 2001). Seminal fluid represents a milieu enabling spermatozoa to break the ovum membrane and suppressing its immune response and, at the same time, protecting male germ cells from infection. 7-OH-DHEA may be one of the components maintaining the tender balance between immuno-protective and immunosuppresive effects.

In conclusion, 7-hydroxylated metabolites of DHEA and pregnenolone pro-duced in cells may act directly in an autocrine manner (e.g. in various brain cells), or by a paracrine mechanism on the neighboring cells (see Lafaye *et al.*, 1999), or on remote targets to which they are transported by blood. The latter fact raises the question of their circulating levels.

PHYSIOLOGICAL LEVELS OF 7-OH-DHEA ISOMERS AND AGEING

Since the pioneering work of Skinner *et al.* (1977), who developed the first, unspe-cific RIA of 7α-OH-DHEA, it has been known that 7-OH-DHEA is present in near nanomolar concentration in human blood. No data have been available, however, on what were their changes during the human life and under various physiological and pathological conditions. Recently we have developed specific immunoassay methods for determination of both 7-OH-DHEA isomers in body fluid, using specific antisera constructed against haptenes prepared via the C_{19} group of the steroid molecule (Lapcík *et al.*, 1998, 1999). This enabled us to measure both 7-OH-DHEA isomers in representative groups of subjects of both sexes during the life, taking advantage of the samples collected during screening of the iodine deficiency in our country.

Using the polynomial regression of the 4th or 5th degree, the dependence of 7α-OH-DHEA levels on age in 172 women aged 10–72 years (A) and 217 men aged 10–91 years (B) was evaluated. The results are shown on Figure 2.2. The central bold curves represent the mean prediction with the 95% confidence interval (the thin lines closer to the mean prediction curve) and the 95% confidence inter-vals of prediction, respectively (the thinnest line more remote from the central line). In contrast to men, where a distinct decline with age occurred, two local maxima have been revealed round the age 22 and 53, respectively, in females. The analo-gous results for 7β-OH-DHEA are shown on Figure 2.3. The curve reflecting the dependence of 7β-OH-isomer on age in women is almost identical with that for 7α-OH-DHEA, in agreement with excellent correlation of the isomers with each other. Figures 2.4 and 2.5 demonstrate this correlation for females and males, respectively. Due to skewed non-Gaussian data distribution in both dimensions, the values were transformed to minimum skewness prior correlation. The princi-pal axis (the bold straight line) and the 95% confidence elipsoid obtained (part A of the Figures) were retransformed to the original scale (part B). It may be noticed that the polynomial curves of 7-OH-DHEA levels reflecting dependence on age in

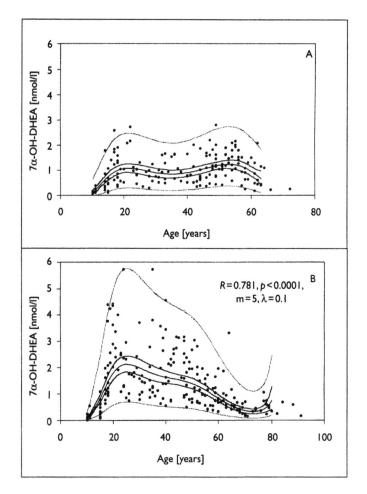

Figure 2.2 Dependence of 7α-OH-DHEA levels on age in healthy subjects. The way of calculation of polynomial regression, mathematical transformation of the data and symbols used are the same as in Figure 2.1. A (upper panel): females, B (lower panel): males.

both sexes resemble strongly that of unconjugated DHEA, but in the case of DHEA the second maximum in women precedes that of both 7-OH-DHEA isomers: while the former peaked round the age 42 years (i.e. at the premeno-pausal period), the latter occurred about 10 years later. Figure 2.6 shows the age dependence of DHEA and its 7-hydroxymetabolites according to age groups divided by 5-years intervals. The circles with horizontal bars represent mean group values with 95% confidence intervals. The data were treated by power transform-ation to minimum skewness prior testing and the results were re-transformed to the original scale. As confirmed by two-way ANOVA with sex and age group as the first and second factor, respectively, both the sex and age differences were highly significant ($p < 0.0001$).

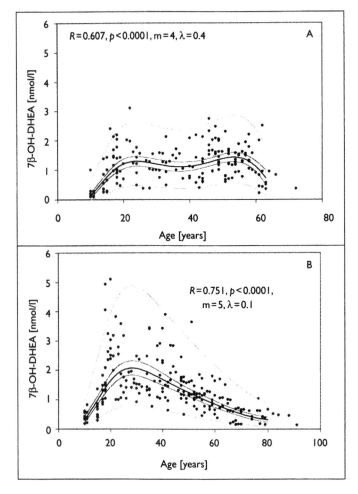

Figure 2.3 Dependence of 7β-OH-DHEA levels on age in healthy subjects. The way of calculation of polynomial regression, mathematical transformation of the data and symbols used are the same as in Figure 2.1. A (upper panel): females, B (lower panel): males.

7-OXO-DHEA

3β-Hydroxy-5-androstene-7,17-dione (7-oxo-DHEA) was isolated from a body fluid as early as in 1954 (Fukushima *et al.*, 1954) and later it was established as a natural constituent of human plasma and urine. Methods have been developed for quantification of 7-oxo-DHEA and its sulfate. For further information see the recent HPLC method of Marwah *et al.* (1999) and the literature therein. Since liver was believed to be the main site of formation of 7-oxygenated 5-ene-steroids, we studied the interconversion of 7α- to 7β-OH-DHEA and *vice versa*, which proceeded via 7-oxo-steroid intermediate (Hampl and Stárka, 1969).

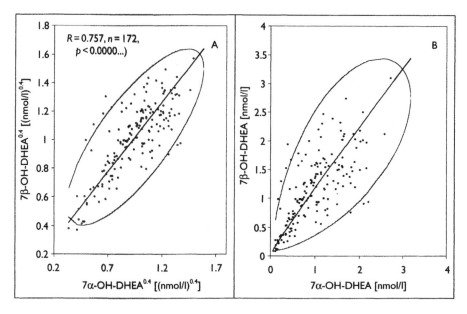

Figure 2.4 Correlation between 7α- and 7β-OH-DHEA levels in 217 healthy males. Due to non-Gaussian distribution, the data were transformed to minimum skewness by power transformation prior to correlation. The central axis (bold straight line) and the 95% confidence elipsoid (A) were re-transformed to the original scale (B). Other symbols are the same as in Figures 2.1–2.3.

The interest for this metabolite increased after discovery of thermogenic properties of DHEA, based on induction of the thermogenic liver mitochondrial enzymes glycerol-3-phosphate dehydrogenase and cytosolic malic enzyme (Su and Lardy, 1991; Bobyleva *et al.*, 1993). Further studies of the same groups revealed that 7-oxo-DHEA was a more potent thermogenic agent than its parent steroids. The term "ergosteroid" was assignated then to 7-oxo-DHEA (Lardy *et al.*, 1995; Bobyleva *et al.*, 1997; Lardy *et al.*, 1998; Reich *et al.*, 1998). Generally, thermogenic effect consists of a shift from oxidative metabolism to heat production, at the expense of formation of energy-rich macroergic phosphates. In these properties 7-oxo-DHEA resembles thyroid hormones. In an experiment with thyroidectomized rats the effect of thyroid hormone deprivation could be restored by administration of 7-oxo-DHEA (Bobyleva *et al.*, 1997). The mechanisms of thermogenesis, however, differ for both classes of compounds. One explanation of thermogenic effect of thyroid hormones is based on stimulation of the gene expression of uncoupling proteins, which uncouple ATP production from mitochondrial respiration and thus decrease the energy metabolism efficiency by its dissipating as heat (Reitman *et al.*, 1999; Schrauwen *et al.*, 1999). Other authors, however, point out that at least in humans this mechanism need not to be the major way by which thyroid hormones act as thermogenic agents (Boivin *et al.*, 2000). Concerning DHEA and its 7-oxo-metabolite, they act at a non-genomic level as inducers of key enzymes of

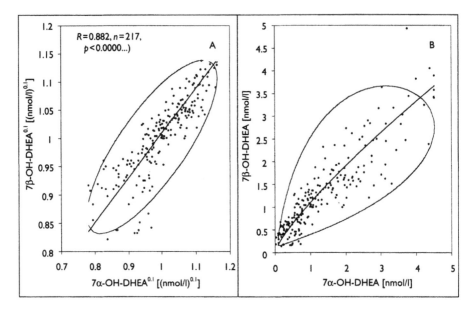

Figure 2.5 Correlation between 7α- and 7β-OH-DHEA levels in 172 healthy females. Due to non-Gaussian distribution, the data were transformed to minimum skewness by power transformation prior correlation. The central axis (bold straight line) and the 95% confidence ellipsoid (A) were re-transformed to the original scale (B). Other symbols are the same as in Figures 2.1–2.3.

mitochondrial and cytoplasmatic oxidative metabolism. The mechanism suggested by Bobyleva *et al.* (1993) and confirmed by further results (Bobyleva *et al.*, 1997; Lardy *et al.*, 1998; Reich *et al.*, 1998) is based on decreased metabolic efficiency via transhydrogenation of cytosolic NADPH into mitochondrial FADH2, with a consequent loss of energy as heat.

It should be noticed that corticoids inhibited gene expression of uncoupling proteins and thus reduced thermogenesis (Moriscot *et al.*, 1993; Arvaniti *et al.*, 1998). 7-oxo-DHEA and DHEA act in opposite ways, though, as demonstrated above, by different mechanisms. This could give an additional example for the anti-glucocorticoid action of 7-oxygenated steroids.

16-HYDROXYLATED METABOLITES OF DHEA

A concurrent metabolic reaction to 7-hydroxylation of DHEA is hydroxylation at carbon 16, resulting in 16α- or 16β-hydroxylated steroids (Baulieu, 1996). 16-Hydroxylated metabolites of DHEA were detected as early as the fifties in umbilical blood of newborns (Fotherby *et al.*, 1957). A series of papers appeared soon after on isolation and identification of various 16α- or 16β-hydroxysteroids in urine of children and adults with various endocrine disorders (Reynolds, 1965;

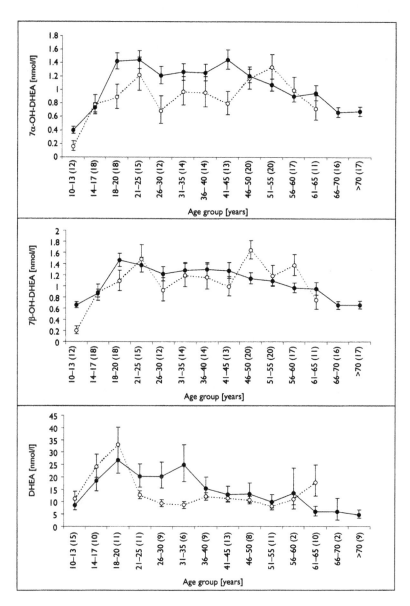

Figure 2.6 Dependence of physiological serum levels of 7α-OH-DHEA, 7β-OH-DHEA and DHEA on age. Full line: – males, dotted line – females. The circles represent group (5 years each) mean values, the horizontal bars 95% confidence intervals calculated by the least significant difference multiple comparison. Overlapping of the confidence intervals denotes statistical insignificance between groups and *vice versa*. The data were treated by power transformation to minimum skewness prior testing and the results were re-transformed to original scale. As confirmed using two-way ANOVA considering the sex and age group as the first and the second factor, respectively, both sex and age differences were highly significant ($p < 0.0001$).

Cleary and Pion, 1968; Shackleton et al., 1968). 16-Hydroxylation of estrogens, androgens and their precursors was at first demonstrated in liver or adrenal cortex, but it occurs also in many other tissues. Enzymes responsible for 16α- or 16β-hydroxylation of steroids are commonly classified as EC 1.14.14.1 enzyme named unspecific monooxygenase, belonging to the cytochrome P450 family. To date not less than 244 genes of which 23 in humans, encoding various proteins and isoenzymes, differing in substrate specificity and tissue localization were described and characterized. With respect to steroid 16α-hydroxylation, two enzymes are of particular importance, namely CYP 3A4 present in human liver microsomes and CYP 1A1 from various tissues including brain (Lacroix et al., 1997).

From the physiological point of view, 16α- and 16β-hydroxylation of steroids is an important enzymatic reaction and a prerequisite for the formation of 16-oxygenated estrogens, of which the most important is estriol (1,3,5(10)-estratriene-3, 16α,17β-triol), the main estrogen in pregnancy. For years its levels have been used as a marker of fetus well being. According to the concept introduced by Diczfalusy in the sixties, and known as the fetoplacental unit, DHEA in a form of sulfate is at first hydroxylated at 16α-positition in fetal liver and then, following passage across the fetoplacental barrier, is hydrolyzed by steroid sulfatase and finally aromatized in placenta (Goodyer and Branchaud, 1981). Generally, insufficient production of placental estrogens in midpregnancy may lead to pathological development of the fetus. The reason of low estrogen production has been searched in decreased DHEAS supply, lowered activity of 16α-hydroxylation, and sulfatase or aromatase deficiency. Several reports brought evidence that sulfatase deficiency is the major cause of estrogen insufficiency (Glass et al., 1998; Wilmot et al., 1988), though it need not be true in all instances as demonstrated by Newby et al. (2000) in Down's syndrome pregnancies, where the main cause was the diminished supply of the fetal DHEAS.

The physiological role of 16α-hydroxydehydroepiandrosterone (16α-OH-DHEA) and 16α-hydroxylated pregnenolone is not known. To our knowledge the only reports on the levels of 16α-OH-DHEA and 16α-hydroxylated pregnenolone and their sulfated derivatives deal with the prenatal or early neonatal period and with women during labor (Furuya et al., 1976; Chang et al., 1976; Shibusawa et al., 1978; Tagawa et al., 1997). The most common material for determination of these steroids was umbilical blood, in which relatively high levels (10 nM) of unconjugated 16α-OH-DHEA were detected. The only paper dealing with physiological levels of 16β-OH-DHEA during the life was published by Sekihara et al. (1976).

Much more effort was devoted to possible physiological and pathophysiological role of 16α-hydroxylated estrogens. It was demonstrated that in mice 16α-hydroxylated estrogens were closely associated with mammary tumors (Bradlow et al., 1986b). Several studies have been undertaken to test whether concentrations of 16α-hydroxylated estrogens and their urinary metabolites do correlate with incidence of breast and endometrial cancer in women, but the results were not unequivocal (Fishman et al., 1984; Bradlow et al., 1986a; Ursin et al., 1997). It has been suggested that cancer promoting effects of natural 16α-hydroxylated estrogens could be due to their easy transformation into covalent adducts to protein macromolecules (as e.g. with estrogen receptors) (Bradlow et al., 1986a; Bhavnani 1998). This reaction does not proceed with 16α-hydroxylated equine estrogens where the B-unsaturated ring is known to reduce carcinogenic properties (Bhavnani, 1998).

In addition, significantly increased serum levels of 16α-hydroxylated estrogens have been found in patients with typical autoimmune diseases such as systemic lupus erythematosus or rheumatioid arthritis (Lahita *et al.*, 1981, 1985; Merril *et al.*, 1996). It is of interest that in these patients low levels of DHEA have been found. (Merril *et al.*, 1996). With respect to recent findings of immunomodulatory and immunoprotective role of DHEA (Kalimi *et al.*, 1994) it would be of interest to find out what the role of the metabolic route leading to the 16α-hydroxylated DHEA metabolite is. In the previous paragraphs, we have shown that 7α- and 7β-hydroxylated steroids may serve as locally active immunoprotective agents, even more potent than DHEA itself in some situations. One may speculate whether a concurrent 16α-hydroxylation may lead to compounds with an opposite effect.

In conclusion, since DHEA and its metabolites are the first targets for 16α-hydroxylation, it seems worth investigating their levels in physiological and pathophysiological conditions. This would contribute to filling the gap in our knowledge. A special respect should be paid to breast and other carcinoma of female reproductive tract, as well as to patients with autoimmune diseases.

The immunoassays developed for these purposes used immunogens prepared from haptens attached to carrier protein through the 3β-position, which is inherent to naturally occurring steroids (Chang *et al.*, 1976; Furuya *et al.*, 1976; Shibusawa *et al.*, 1978; Tagawa *et al.*, 1997). Thus the methods suffered from a poor specificity. Therefore, the first task to complete is to develop a reliable and specific method for determination of 16α-OH-DHEA.

ACKNOWLEDGMENT

This work was supported by the Grant No. 5397-3 of the Internal Grant Agency of the Czech Ministry of Health.

REFERENCES

Akwa, Y., Sananes, N., Gouezou, M., Robel, P., Baulieu, E.E., and Le Goascogne, C. (1993) Astrocytes and neurosteroids: metabolism of pregnenolone and dehydroepiandrosterone. Regulation by cell density. *J. Cell Biol.*, **121**, 135–143.

Arvaniti, K., Ricquier, D., Champigny, O. and Richard, D. (1998) Leptin and corticosterone have opposite effects on food intake and the expression of UCP1 mRNA in brown adipose tissue of lep(ob)lep(ob) mice. *Endocrinology*, **139**, 4000–4003.

Ballabio, A. and Shapiro, L.J. (1995) Steroid sulfatase deficiency and X-linked ichtyosis. In C.R. Shriver, A.L. Beaudet, W.S. Sly and D. Valle (eds) *The metabolic and molecular basis of inherited disease*, McGraw-Hill, New York, pp. 2999–3022.

Baulieu, E.E. (1996) Dehydroepiandrosterone (DHEA): a fountain of youth? *J. Clin. Endocrinol. Metab.*, **81**, 3147–3151.

Baulieu, E.E., Thomas, G., Legrain, S., Lahlou, N., Roger, M., Debuire, B. *et al.* (2000) Dehydroepiandrosterone (DHEA), DHEA sulfate and aging: contribution of the DHEA age study to sociobiomedical issue. *Proc. Natl. Acad. Sci. USA*, **97**, 4279–4284.

Bhavnani, B.R. (1998) Pharmacokinetics and pharmacodynamics of conjugated equine estrogens: chemistry and metabolism. *Proc. Soc. Exp. Biol. Med.*, **217**, 6–16.

Bobyleva, V., Bellei, M., Kneer, N. and Lardy, H. (1997) The effect of ergosteroid 7-oxo-dehydroepiandrosterone on mitochondrial membrane potential: possible relationship to thermogenesis. *Arch. Biochem. Biophys.*, **341**, 122–128.

Bobyleva, V., Kneer, N., Bellei, M., Batelli, D. and Lardy, H.A. (1993) Concerning the mechanism of increased thermogenesis in rats treated with dehydroepiandrosterone. *J. Bioenerg. Biomembr.*, **25**, 313–321.

Boivin, M., Camirand, A., Carli, F., Hoffer, L.J. and Silva, J.E. (2000) Uncoupling protein-2 and -3 messenger ribonucleic acids in adipose tissue and skeletal muscle of healthy males: variability, factors affecting expression and relation to measures of metabolic rate. *J. Clin. Endocrinol. Metab.*, **85**, 1975–1983.

Bradlow, H.L., Hershcopf, R.J. and Fishman, J. (1986a) Oestradiol 16α-hydroxylase: a risk marker for breast cancer. *Cancer Surv.*, **5**, 573–583.

Bradlow, H.L., Hershcopf, R.J., Martucci, C.P. and Fishman, J. (1986b) 16α-Hydroxylation: a possible risk marker for breast cancer. *Ann. N. Y. Acad. Sci.*, **464**, 138–151.

Chang, R.J., Buster, J.E., Blakely, J.L., Okada, D.M., Hobel, C.J., Abraham, G.E. *et al.* (1976) Simultaneous comparison of Δ^5-3β-hydroxysteroid levels in the fetoplacental circulation of normal pregnancy in labor and not in labor. *J. Clin. Endocrinol. Metab.*, **42**, 744–751.

Chmielewski, V., Drupt, F. and Morfin, R. (2000) Dexamethasone-induced apoptosis of mouse thymocytes: Prevention by native 7α-hydroxysteroids. *Immunol. Cell Biol.*, **78**, 238–246.

Cleary, R.E. and Pion, R.J. (1968) Urinary excretion of 16α-hydroxydehydroepiandroster-one and 16-ketoandrostenediol during the early neonatal period. *J. Clin. Endocrinol. Metab.*, **28**, 372–378.

Compagnone, N.A., Salido, E., Shapiro, L.J. and Mellon, S.H. (1997) Expression of steroid sulfatase during embryogenesis. *Endocrinology*, **138**, 4768–4773.

Cuevas-Covarrubias, S.A., Juarez-Oropeza, M.A., Miranda-Zamora, R. and Diaz-Zagoya, J.C. (1993) Comparative analysis of human steroid sulfatase activity in prepubertal and postpubertal males and females. *Biochem. Mol. Biol. Int.*, **30**, 691–695.

Doostzadeh, J. and Morfin, R. (1996) Studies of the enzyme complex responsible for pregnenolone and dehydroepiandrosterone 7α-hydroxylation in mouse tissues. *Steroids*, **61**, 613–620.

Fishman, J., Schneider, J., Hershcopf, R.J. and Bradlow, H.L. (1984) Increased estrogen-16α-hydroxylase activity in women with breast and endometrial cancer. *J. Steroid Biochem.*, **20**, 1077–1081.

Fotherby, K., Colás, A., Atherden, S.M. and Marrian, G.F. (1957) The isolation of 16α-hydroxydehydroepiandrosterone (3β,16α-dihydroxyandrost-5-en-17-one) from the urine of normal men. *Biochem. J.*, **66**, 664–669.

Fukushima, D.K., Kemp, A.D. and Schneider, R. (1954) Studies in steroid metabolism XXV. Isolation and characterization of new urinary steroids. *J. Biol. Chem.*, **210**, 129–137.

Furuya, K., Yoshida, T., Takagi, S., Kanbegawa, A. and Yamashita, H. (1976) Radioimmu-noassay of 16α-hydroxy-dehydroepiandrosterone and its sulfate. *Steroids*, **27**, 797–812.

Glass, I.A., Lam, R.C., Chang, T., Roitman, E., Shapiro, L.J. and Shackleton, C.H. (1998) Steroid sulphatase deficiency is the major cause of extremely low oestriol production in mid-pregnancy: a urinary assay for the discrimination of steroid sulphatase deficiency from other causes. *Prenat. Diagn.*, **18**, 789–800.

Goodyear, C.G. and Branchaud, C.L. (1981) Regulation of hormone production in the human feto-placental unit. *Ciba Found. Symp.*, **86**, 89–123.

Hampl, R., Hill, M., Šterzl, I. and Stárka, L. (2001) Immunomodulatory 7-hydroxylated metabolites of dehydroepiandrosterone are present in human semen. *J. Steroid Biochem. Molec. Biol.*, In the press.

Hampl, R., Morfin, R. and Stárka, L. (1997) 7-Hydroxylated derivatives of dehydro-epiandrosterone: what are they good for? *Endocrine Regul.*, **31**, 211–218.

Hampl, R. and Stárka, L. (1969) Epimerisation of naturally occurring C-19 steroid allylic alcohols by rat liver preparations. *Europ. J. Steroids (later J. Steroid Biochem.)*, **1**, 47–56.

Kalimi, M., Shafagoj, Y., Loria, R., Padgett, D. and Regelson, W. (1994) Anti-glucocorticoid effects of dehydroepiandrosterone (DHEA). *Molec. Cell. Biochem.*, **131**, 99–104.

Khalil, M.W., Strutt, B., Vachon, D., and Killinger, D.W. (1994) Effect of dexametnasone and cytochrome P450 inhibitors on the formation of 7alpha-hydroxydehydroepiandro-sterone by human adipose stromal cells. *J. Steroid Biochem. Mol. Biol.*, **48**, 545–552.

Lacroix, D., Sonnier, M., Moncion, A., Cheron, G. and Cresteil, T. (1997) Expression of CYP3A in the human liver – evidence that the shift between CYP3A7 and CYP3A4 occurs immediately after birth. *Europ. J. Biochem.*, **247**, 625–634.

Lafaye, P., Chmielewski, V., Nato, F., Mazié, J.-C. and Morfin, R. (1999) The 7α-hydroxy-steroids produced in human tonsils enhance the immune response to tetanus toxoid and *Bordetella pertusis* antigens. *Biochim. Biophys. Acta*, **1472**, 222–231.

Lahita, R.G., Bradlow, H.L., Kunkel, H.G. and Fishman, J. (1981) Increased 16α-hydroxy-lation of estradiol in systemic *lupus erythematosus*. *J. Clin. Endocrinol. Metab.*, **53**, 174–178.

Lahita, R.G., Bucala, R., Bradlow, H.L. and Fishman, J. (1985) Determination of 16α-hydroxy-estrone by radioimmunoassay in systemic *lupus erythematosus*. *Arthritis Rheum.*, **28**, 1122–1127.

Lapcík, O., Hampl, R., Hill, M., Bicíková, M. and Stárka, L. (1998) Immunoassay of 7-hydroxysteroids: 1. Radioimmunoassay of 7β-hydroxy-dehydroepiandrosterone. *J. Steroid Biochem. Molec. Biol.*, **67**, 439–445.

Lapcík, O., Hampl, R., Hill, M. and Stárka, L. (1999) Immunoassay of 7-hydroxysteroids: 2. Radioimmunoassay of 7α-hydroxy-dehydroepiandrosterone. *J. Steroid Biochem. Molec. Biol.*, **71**, 231–237.

Lardy, H., Kneer, N., Wei, Y., Partridge, B. and Marwah, P. (1998) Ergosteroids II. Biologically active metabolites and synthetic derivatives of dehydroepiandrosterone. *Steroids*, **63**, 158–165.

Lardy, H., Partridge, B., Kneer, N. and Wei, Y. (1995) Ergosteroids: induction of thermogenic enzymes in liver of rats treated with steroids derived from dehydroepiandrosterone. *Proc. Natl. Acad. Sci. USA*, **92**, 6617–6619.

Longcope, C. (1996) Dehydroepiandrosterone metabolism. *J. Endocrinology*, **150**, S125–S127.

Marwah, A., Marwah, P. and Lardy, H. (1999) Development and validation of a high-performance liquid chromatography assay for the quantitative determination of 7-oxo-dehydroepiandrosterone-3β-sulfate in human plasma. *J. Chromatogr. B. Biomed. Sci. Appl.*, **721**, 197–205.

Merril, J.T., Dinu, A.R. and Lahita, R.G. (1996) Autoimmunity: the female connection. *Medscape Womens Health*, **1**, 5.

Milone, A., Delfino, M., Piccirillo, A., Illiano, G.M., Aloj, S.M. and Bifulco, M. (1991) Increased levels of DHEAS in serum of patients with X-linked ichtyosis. *J. Inherit. Metab. Dis.*, **14**, 96–104.

Miranda-Duarte, A., Valdes-Flores, M., Miranda-Zamora, R., Diaz-Zagoya, J.C., Kofman-Alfaro, S.H. and Cuevas-Covarrubias, S.A. (1999) Steroid sulfatase activity in leucocytes: a comparative study in 45,X; 46Xi(Xq) and carriers of steroid sulfatase deficiency. *Biochem. Mol. Biol. Int.*, **47**, 137–142.

Morfin, R. and Courchay, G. (1994) Pregnenolone and dehydroepiandrosterone as precursors of native 7-hydroxylated metabolites which increase the immune response in mice. *J. Steroid Biochem. Molec. Biol.*, **50**, 91–100.

Moriscot, A., Rabelo, R. and Bianco, A.C. (1993) Corticosterone inhibits uncoupling protein gene expression in brown adipose tissue. *Am. J. Physiol.*, **265**, (1 Pt 1), E81–E87.

Newby, D., Aitken, D.A., Howatson, A.G. and Connor, J.M. (2000) Placental synthesis of oestriol in Down's syndrome pregnancies. *Placenta*, **21**, 263–267.

Padgett, D.A. and Loria, R.M. (1994) *In vitro* potentiation of lymphocyte activation by dehydroepiandrosterone, androstenediol and androstenetriol. *J. Immunol.*, **153**, 1544–1552.

Reich, I.L., Lardy, H., Wei, Y., Marwah, P., Kneer, P., Powell, D.R. *et al.* (1998) Ergosteroids III. Syntheses and biological activity of secosteroids related to dehydroepiandrosterone. *Steroids*, **63**, 542–553.

Reitman, M.L., He, Y. and Gong, D.W. (1999) Thyroid hormone and other regulators of uncoupling proteins. *Int. J. Obes. Relat. Metab. Disord.*, **23**, Suppl. 6, S56–S59.

Reynolds, J.W. (1965) Excretion of two Δ^5-3β-OH, 16α-hydroxysteroids by normal infants and children. *J. Clin. Endocrinol. Metab.*, **25**, 416–423.

Rose, K.A., Stapleton, G., Dott, K., Kieny, M.P., Best, R., Schwarz, M. *et al.* (1997) Cyp7b, a novel brain cytochrome P450, catalyzes the synthesis of neurosteroids 7α- hydroxy-dehydroepiandrosterone and 7α-hydroxypregnenolone. *Proc. Nat. Acad. Sci. USA*, **94**, 4925–4930.

Rupprecht, H. and Holsboer, F. (1999) Neuroactive steroids: mechanism of action and neuropsychopharmacological perspectives. *Trends Neurosci.*, **22**, 410–416.

Schrauwen, P., Walder, K. and Ravussin, E. (1999) Human uncoupling proteins and obesity. *Obes. Res.*, **7**, 97–105.

Sekihara, H., Sennett, J.A., Liddle, G.W., McKenna, T.J. and Yarbro, L.R. (1976) Plasma 16β-hydroxydehydroepiandrosterone in normal and pathological conditions in man. *J. Clin. Endocrinol. Metab.*, **43**, 1078–1084.

Shackleton, C., Kelly, R.W., Adhikary, P.M., Brooks, R.A., Harkness, R.A., Sykes, P.J. *et al.* (1968) The identification and measurement of a new steroid 16β-hydroxydehydroepiandrosterone in infant urine. *Steroids*, **12**, 705–716.

Shibusawa, H., Sano, Y., Yamamoto, T., Kambegawa, A., Ohkawa, T., Satoh, N., Okinaga, S. *et al.* (1978) A radioimmunoassay of serum 16α-hydroxypregnenolone with specific antiserum. *Endocrinol. Jpn.*, **25**, 185–189.

Skinner, S.J.M., Tobler, C.J.P. and Couch, R.A.F. (1977) A radioimmunoassay for 7α-hydroxydehydroepiandrosterone in human plasma. *Steroids*, **30**, 315–330.

Stapleton, G., Steel, M., Richardson, M., Mason, J.O., Rose, K.A., Morris, R.G.M. *et al.* (1995) A novel cytochrome P450 expressed primarily in brain. *J. Biol. Chem.*, **270**, 29739–29745.

Stárka, L., Hill, M., Hampl, R., Morfin, R., Malewiak, M.-I., Kolena, J. *et al.* (1998) On the mechanism of antiglucocorticoid action of 7α-hydroxy-dehydroepiandrosterone: effect on DNA binding of dexamethasone-labeled glucocorticoid receptor and on membrane fluidity. *Coll. Czech. Chem. Commun.*, **63**, 1683–1698.

Stárka, L., Šulcová, J. and Šilink, K. (1962) Die Harnausscheidung des 7-hydroxydehydroepiandrosteronsulfates. *Clin. Chim. Acta*, **7**, 309–316.

Su, C.Y. and Lardy, H. (1991) Induction of hepatic mitochondrial glycerophosphate dehydrogenase in rats by dehydroepiandrosterone. *J. Biochem. (Tokyo)*, **110**, 207–213.

Suitters, A.J., Shaw, S., Wales, M.R., Porter, J.P., Leonard, J., Wootger, R. *et al.* (1999) Immune enhancing effects of dehydroepiandrosterone sulphate and the role of steroid sulphatase. *Immunology*, **91**, 314–321.

Šulcová, J., Hill, M., Hampl, R. and Stárka, L. (1997) Age and sex related differences in serum levels of unconjugated dehydroepiandrosterone and its sulphate in normal subjects. *J. Endocrinol.*, **154**, 57–62.

Šulcová, J., Novák, J., Jirásek, J.E. and Stárka, L. (1968) Distribution of dehydroepiandrosterone 7-hydroxylating activity in human tissues. *Endocr. Exper.*, **2**, 157–172.

Šulcová, J. and Stárka, L. (1968) Characterization of microsomal dehydroepiandrosterone 7-hydroxylase from rat liver. *Steroids*, **12**, 113–126.

Tagawa, N., Kusuda, S., and Kobayashi, Y. (1997) C16 hydroxylation of 3beta-hydroxy-delta5-steroids during the early neonatal period. *Biol. Pharm. Bull.*, **20**, 1295–1299.

Ursin, G., London, S., Stanczyk, F.Z., Gentzschein, E., Paganini-Hill, A., Ross, R.K. *et al.* (1997) A pilot study of urinary estrogen metabolites (16α-OHE1 and 2-OHE1). *Environ. Health Perspect.*, **105**, Suppl. 3, 601–605.

Ursin, G., London, S., Stanczyk, F.Z., Gentzschein, E., Paganini-Hill, A., Ross, R.K. *et al.* (1999) Urinary 2-hydroxysterone ratio and risk of breast cancer in postmenopausal women. *J. Natl. Cancer Inst.*, **91**, 1067–1072.

Wilmot, R.L., Mawson, R.J. and Oakey, R.E. (1988) Antenatal detection of placental steroid sulphatase deficiency by measurement of urinary 16α-hydroxydehydroepiandrosterone sulphate. *Ann. Clin. Biochem.*, **25**, 155–161.

Wolf, O.T. and Kirschbaum, C. (1999) Actions of dehydroepiandrosterone and its sulfate in the central nervous system: effect on cognition and emotion in animals and humans. *Brain Research Review*, **30**, 264–288.

Zumoff, B., Rosenfeld, R.S., Strain, G.W., Levin, J. and Fukushima, D.K. (1980) Sex differences in the twenty-four hour mean plasma concentrations of dehydroisoandrosterone (DHA) and dehydroisoandrosterone sulfate (DHAS) and the DHA to DHAS ratio in normal adults. *J. Clin. Endocrinol. Metab.*, **51**, 330–333.

Chapter 3

Postulated roles of DHEA in the decline of neural function with age

L.-Y. Ngai and J. Herbert

INTRODUCTION

Dehydroepiandrosterone (DHEA) has been known since the 1930s. Only in the last 10 years has it attracted much attention outside those studying the feto-placental unit (in which it has long been known to play a role: Diczfalusy (1984)) or as a clinical aid in the diagnosis of some ovarian disorders. Recently, however, increasingly large numbers of (mostly middle-aged) people have been taking DHEA (usually in an unregulated manner), and it has acquired a popular reputation as a rejuvenating or anti-ageing therapy. Such compounds have a long history: the elixir of youth has been a recurring feature of many a nostrum, and the foundation of many doubtful reputations over the centuries. Is DHEA just another passing fad, destined to join the monkey glands and rhino horns of yesteryear? Or is there some empirical basis for the current interest in this steroid? Here we focus on the current evidence that DHEA has a role to play in the detrimental effects of age on the brain.

THE LIFE HISTORY OF DHEA

DHEA has a rather extraordinary trajectory over the lifespan, distinct from any other steroid. During intra-uterine life, the fetal (human) adrenal secretes massive amounts of DHEA (and its sulfate, DHEAS) from an hypertrophied zona fasciculata (the "fetal zone"); this depends on the activity of the enzyme $P450_{cyp17}$, which forms Δ^5 C_{19} steroids such as DHEA, and the absence of 3β-hydroxysteroid dehydrogenase (3βHSD) which converts them to Δ^4 steroids such as cortisol (or corticosterone) (Winter, 1998). Stress increases fetal DHEA even further. It has always been assumed that the major function of fetal DHEA was to supply the placenta (which lacks $P450_{cyp17}$) with its required precursor for estrogens (in particular, the 16-OH compound estriol). Whilst this is undoubtedly true, it may not be the only role for fetal DHEA. Relative or complete absence of DHEA in fetal life does not prevent continuation of pregnancy, or formation of at least some estriol; and DHEA itself may affect the development of the fetal brain – by changing the growth of dendrites (Compagnone and Mellon, 1998). Whether there is an interaction between glucocorticoids and DHEA in fetal life, as there may be in adults (see below) awaits further evidence, but it may be a topic well worth exploring.

Table 3.1 Mean (ng/ml±SD) levels of cortisol and DHEA and the molar cortisol/ DHEA at two time points in the saliva of 116 adult women (23–58 years) (from Harris *et al.*, 2000)

	08.00 h	20.00 h
Cortisol	3.54±1.97	0.77±1.59
DHEA	0.32±0.15	0.19±0.09
Molar cortisol/DHEA ratio	10.53	3.84

At birth, there is an abrupt fall in DHEA, unlike cortisol, which continues to be secreted. Levels of DHEA continue to be very low until about 8 years in man, when there is a progressive rise ("adrenarche") (Parker, 1991, 1993). DHEA, and its sulfate DHEAS, continue to increase during puberty so that, by young adulthood, DHEAS levels are at least 10 times higher than cortisol (often even higher), whereas DHEA is about 10% of DHEAS, with values similar to that of testosterone (in males). It should be noted that these high DHEA/S levels are characteristic of primates, and are not observed in laboratory rodents. Levels peak in the early 20s, and herald a further extraordinary phase in the saga of DHEA. From early adulthood (2nd–3rd decade), DHEA/S levels decline remorselessly, at about 10–15% per decade (the rate is individually rather variable) so that by age 70, levels are about 20% or less of those at the peak (Orentreich *et al.*, 1984, 1992). Though other steroids decrease with age (some precipitously, such as the gonadal steroids in women at the menopause), none shows this pattern of progressive, sustained and marked decline in both sexes. It should be noted that cortisol neither increases during adrenarche and puberty nor does it decline with age (in fact, there may be increased secretion in the elderly, particularly during the evening: see below). It is clear, therefore, that cortisol and DHEA have different life histories.

They also have different diurnal rhythms. A characteristic feature of cortisol is its rhythmic change during the day in man; levels in the blood are about three times as high in the morning as in the evening; in the saliva (which measures "free" cortisol), this rhythm is accentuated, so that its amplitude is about 8–10-fold (Table 3.1). DHEA, by contrast, has a much less marked rhythm (salivary levels vary about 2-fold), whereas DHEAS (which has a very long half-life) has little discernible rhythm. The two steroids also respond differently to stress in the adult: cortisol, the archetypical "stress" hormone, increases rapidly in response to stressors (e.g. illness) whereas DHEA decreases. These findings show that even though ACTH can increase the secretion of both cortisol and DHEA, there must be distinct control mechanisms for the two steroids. This becomes all the more significant in view of the possibility, discussed below, that DHEA moderates the actions of cortisol, and that, therefore, the relative concentrations of the two hormones may be important for cerebral function. But there is a further complication: DHEA may be made in the brain.

DHEA AS A NEUROSTEROID

DHEA and its sulfate conjugate DHEAS, are both members of a family of steroid hormones, classed as "neurosteroids" (Baulieu, 1998). This term was applied to those steroids that are synthesized in the brain, either *de novo* from cholesterol or by *in situ* metabolism of steroid precursors. DHEA/S is found in the brain of adult male rats. This finding is rather surprising, particularly as the rodent's adrenals do not secrete appreciable amounts of the steroid. In fact, concentrations of DHEA/S are at the limit of sensitivity when measured in the plasma by radio-immunoassay (Ngai *et al.*, unpublished data; Robel and Baulieu, 1994). In contrast, the levels of DHEA/S are found to be higher in the brain when compared to plasma levels (between 0.2–5 ng/g compared to 0.1–0.3 ng/g) (Robel and Baulieu, 1994). To demonstrate that DHEA/S found in the brain is independent of peripheral steroidogenesis (and is a true neurosteroid), male rats were castrated and adrenalectomized. DHEA/S concentrations were unchanged and persisted in the brain for several weeks, whereas levels of testosterone disappeared in the brain (Corpéchot *et al.*, 1981, 1983). To demonstrate that brain DHEA does not result from accumulation or retention of DHEA from peripheral sources, radio-active DHEA was injected into male intact rats, and levels were then measured. The concentrations of radioactive DHEA were indeed found to be higher in the brain than in plasma, but they did not approach the concentration ratios observed with levels of endogenous steroid in the brain and plasma. Further-more, radioactive DHEA was retained in the brain for the same duration as plasma, and was then undetectable 20 hours after injection (Corpéchot *et al.*, 1983). Consistent wih the hypothesis that DHEA/S is a neurosteroid, a circadian rhythm of DHEA/S in the brain has been found to be unrelated to blood rhythms (Robel *et al.*, 1987), thus suggesting synthesis of DHEA/S is independent of peripheral sources.

However, there are problems with the idea that the adult brain makes DHEA/S. There is no direct evidence for the synthesis of DHEA/S in the adult brain of the rat. Incubation of its radioactive precursor [^{3}H]Pregnenolone (PREG) with either brain slices, or mixed primary cultures of rat and mouse embryo brains, failed to produce [^{3}H]DHEA (Robel and Baulieu, 1995) and the enzyme responsible for conversion of PREG to DHEA, P450$_{cyp17}$ has not been demonstrated in the adult brain by either immunohistochemistry, RNase protection assays or RT-PCR (Mellon and Deschepper, 1993). Therefore, the biosynthesis pathway of DHEA in the brain remains controversial. It is possible that the fetal brain may synthesize DHEA, since P450$_{cyp17}$ has been reported in embryos (Compagnone and Mellon, 1998; Zwain and Yen, 1999). DHEA/S are also found abundantly in human and primate brain, which may result from adrenal and local synthesis of DHEA/S (Robel *et al.*, 1987). However the significance of these findings remains question-able. Are these brain levels functional, do they contribute significantly to neuronal actions of blood-borne steroids? Many observers remain undecided on this issue; since the brain (CSF) levels of DHEA in man are about 5% of those in the blood (Guazzo *et al.*, 1996), it is clear that, in primates (in whom, as we have seen, blood levels are high) the circulation is a major source of cerebral DHEA. Nevertheless, DHEA/S, previously considered to be only an intermediary steroid in the biosyn-thetic pathway of sex steroid hormones (e.g. androgens and estrogens), may be an important steroid hormone in its own right. In particular, there is accumulating evidence that DHEA/S has direct actions on neural function. If this is so, then the progressive decline in DHEA levels with age gains added meaning.

MEASURING DHEA IN THE BLOOD, BRAIN AND SALIVA

In common with other steroids, assessing the role of DHEA on brain function depends on an accurate estimate of its level in various compartments of the body. This, however, is not a simple procedure: many conclusions about DHEA (and other steroids such as cortisol) are based on inadequate knowledge of the basic requirements for reliable measures. There are four fundamental parameters: the assay, the sampling procedure, the time of sampling, and the condition of the subject.

Assays for DHEA

Nearly all estimates for DHEA are based on radioimmunoassays (RIAs), though there are occasional reports of mass-spectrographic estimations (widely regarded as the "gold standard"). Other steroids (including cortisol) are also measured using ELISA procedures, which are often more sensitive than RIAs, but there does not seem to be an adequate ELISA (so far) for DHEA. The theory and practice of RIAs have been extensively reviewed, and a further extended account is unnecessary here. RIAs need to be validated: that is, their sensitivity, specificity and accuracy must be defined by procedures now well established. In the case of DHEA in man, there is a complication: the presence, in the blood, of levels of DHEAS that are 10 times or more those of DHEA. The assumption that DHEAS is a "metabolite" of DHEA, and thus biologically inactive, is almost certainly incorrect. This implies that levels of DHEAS are themselves of interest. However, it cannot be assumed that DHEA and DHEAS are equivalent (see below) so that separate estimates are needed for a complete picture of steroid levels. Many antibodies for DHEA cross-react, to some extent, with DHEAS; even if this is small, the presence of so much DHEAS will ensure inaccuracy unless steps are taken to remove most or all of the DHEAS before assay. Solvent extraction is one method: e.g. hexane/ether extracts almost all DHEA, but leaves most DHEAS behind; ethanol does the opposite. After such an extraction, use of an antibody with a small cross-reactivity becomes feasible.

DHEA in the blood

Another significant issue is the tissue assayed. Most measures of DHEA/S are carried out on blood. It is important to recognize that steroids such as DHEA probably have no function in the blood: what one measures, therefore, in the blood, is the assumed delivery load to hormone-responsive tissues. As we shall see, the relative amounts of DHEA and DHEAS in the blood are very different from those in other tissues (because of differential passage across membranes); levels in the latter may be rather closer to those to which targets are actually exposed. Furthermore, in the case of some other steroids, the presence in the blood of steroid-binding globulins (e.g. SHBG, CBG) results in there being two species of steroid: one bound (and thus less available to target tissues) and the other free (and thus available for diffusion across membranes). There is no known globulin binding protein for DHEA (or DHEAS), though there is the paradox (see below) that levels in other tissues seem to indicate some limiting process for this steroid that resembles that observed for other steroids (e.g. cortisol) that do have a blood binding protein. There is increasing interest in the specific roles of "free" (e.g. unbound) steroids (e.g. testosterone in ageing males, estradiol in post-menopausal women), so this feature of DHEA urgently needs clarification.

DHEA in the brain

Since we are interested, in this chapter, in the effects of DHEA/S on the brain, the passage of these steroids into the brain from the blood becomes a critical issue. Levels of steroids in the brain are regulated by two processes: entry, which depends on diffusion (or transport, in the case of some substances) into the brain, and clearance, which depends both on metabolism in the brain and passage out of the brain into the venous drainage. Before considering DHEA, we should briefly review the evidence for cortisol, both because information for this steroid is more complete than for DHEA, and because, as we shall see, the relative amounts of cortisol and DHEA in the brain may be a critically important parameter.

In the case of cortisol, entry is limited by globulin-binding in the blood – the role of binding to albumin, which has a lower affinity but much greater capacity, is still discussed. Since about 5% of cortisol is "free" under basal conditions – that is, when cortisol is at its diurnal nadir – it is no surprise that simultaneous measures of cortisol in blood and CSF show levels in the latter to be about 5% of those in the blood (Table 3.2). However, two caveats are important: if cortisol levels in the blood rise for any reason (e.g. in response to the diurnal rhythm or following stress) then the binding globulin (CBG) may become saturated. This will result is a much greater proportion of "free" cortisol, and hence additional (non-linear) entry into the brain. Under such circumstances, measuring the total amount of cortisol in the blood will give a very inaccurate estimate of the amounts actually reaching the brain. There is rather little information on the clearance of cortisol from the brain; what there is suggests that it may be slower than from the blood (Martensz et al., 1983), another significant factor in determining CSF levels and hence those in the brain's extracellular compartment.

Levels of DHEA in the CSF of humans are also about 5% of those in the blood, suggesting some limiting factor in the passage of this steroid into the cerebral compartment (Table 3.2). There is no data on the clearance of DHEA from the brain in man. The paradox of this finding, in the absence of a demonstrable blood binding protein for DHEA, has already been mentioned. We do not know whether

Table 3.2 Mean levels of cortisol, DHEA, and DHEAS in the blood and CSF of 62 subjects (ages 3–85) (from Guazzo et al., 1996)

	Cortisol	DHEA	DHEAS
Blood (ng/ml)	101.7	4.1	846.3
CSF	5.7	0.15	1.33
CSF/blood ratio (%)	6.6	6.4	0.21

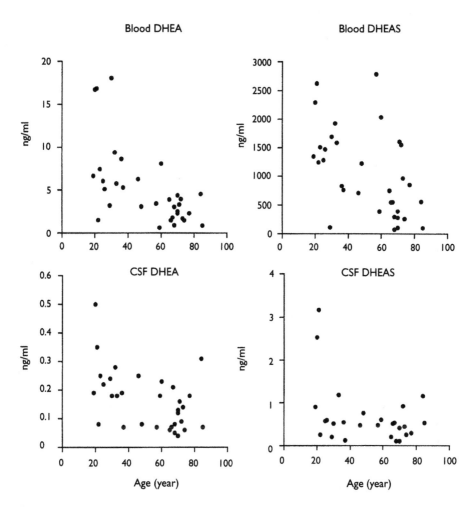

Figure 3.1 Age-related changes in DHEA and DHEAS in the blood and CSF of humans (from Guazzo *et al.*, 1996).

changing the levels of DHEA within an individual will alter the relationship between blood and brain levels of DHEA (as it is known to do for cortisol). What we do know is that levels of DHEA in the blood and CSF are quite closely correlated, and that DHEA levels in the CSF decline with age in man as they do in the blood (Figure 3.1). Thus there is no doubt that the ageing brain is exposed to less DHEA. The critical question, of course, is whether this has any physiological or pathological significance.

It is important to know that the passage of DHEAS into the brain is proportionately much less than for DHEA. Levels of DHEAS in the CSF of man are only about 0.2% of those in the blood (Table 3.2). The reasons for this are not altogether clear, though DHEAS – as would be expected for a conjugate – is much less soluble in lipid than DHEA (i.e. is more polar) and this is likely to be an important factor; passage across membranes will be much less than for DHEA. However, because blood levels of DHEAS are so much higher than DHEA, even this relatively poorer penetration results in DHEAS values for the CSF which exceed those of DHEA. Levels of DHEAS in the CSF that are thus about 2–3 times those of DHEA, compared to the blood, in which DHEAS is more than 100 times the levels of DHEA (Table 3.2). It is worth mentioning again that it cannot be assumed that DHEAS is not active in the brain: in fact, there is experimental evidence suggesting the opposite (see below). DHEAS levels in the CSF also decline with age, as they do in the blood (Figure 3.1).

DHEA in the saliva

Measuring steroid levels in the saliva has become increasingly popular during the past decade or two, for good reasons. Collecting saliva is easy, can be done by the subjects themselves, does not require the assistance of skilled personnel, is non-invasive, and can thus be used to survey steroid levels in large samples of people in, for example, the community. This is very difficult to achieve using blood, though methods based on the collection of finger-prick samples collected onto filter paper have helped. Some simple precautions are needed to ensure accurate results from salivary samples. Using artificial aids to salivation (e.g. lemon juice) may interfere with assays; collecting saliva by using cotton–wool swabs (or similar devices) may do likewise; the best method is to allow the subjects to dribble saliva directly into plastic tubes. Since blood levels are so much higher than those in the saliva it is essential to ensure that there is no bleeding from the teeth or gums. Subjects should be forbidden to brush their teeth before collection, and samples should be checked for blood contamination (e.g. by using either hemoglobin or other blood-specific measures). Rinsing out the mouth with clean water before collection reduces the amount of food particles and other debris in the samples. Subjects of all ages (including children) have no difficulty in providing saliva. Steroids such as cortisol and DHEA are very stable in saliva, which can even be sent by mail from the point of collection to the assay laboratory. Stored at –20 °C, samples can be kept for years if need be.

Steroid levels in saliva have little importance in their own right, but are only interesting as a guide to those in the blood. Simultaneous sampling of blood and saliva shows this to be the case: for cortisol and DHEA, levels in the saliva are about 5% of those in the blood (i.e. they seem to represent the "free" fraction, though the caveat in the case of DHEA should be recalled). DHEAS in the saliva is about 0.2% of blood levels. There is a good correlation between blood levels of DHEA and those in the their saliva in different individuals (Goodyer et al., 1996). Levels in the saliva also decline with age, as expected, since they reflect those in the blood. Steroids are secreted as pulses: rapid changes in the blood levels of cortisol are reflected in similar pulses in the saliva with a time delay of about 15 minutes (del Campo et al., 2000). Whether this is true for DHEA has not been adequately tested.

It will be obvious that salivary levels of DHEA (as well as those of DHEAS and cortisol) resemble those in the CSF. Simultaneous measures of DHEA (or the other steroids) in saliva and CSF have not been reported, so this remains an inference, though it seems likely that the same membrane-dependent factors affect transport into both compartments. If this is proved to be the case, then salivary measures may actually tell us more about levels in the brain than assays of total levels of steroids in the blood.

DHEA: timing the sample

The secretion of DHEA, like other adrenal steroids, is heavily dependent upon input from the environment. It is therefore essential to define the time of day, and the physiological and psychological state of the subject, if an assay result is to have any meaning. It is remarkable how frequently these conditions are not met. Blood levels of cortisol, the best studied of the adrenal steroids, varies in man about 3-fold during the day (maximum at about 8.00 a.m. minimum at about 8.00 p.m.). Levels in the saliva, however, show an accentuated rhythm; levels in the morning are about 8–10-fold those in the evening, probably because the binding capacity of CBG is exceeded during the morning peak. DHEA shows a similar, but lesser, diurnal rhythm in the saliva; DHEAS (which has a much longer half-life in the blood) has little discernible daily rhythm in the saliva. It is important to note that, because the diurnal rhythms for cortisol and DHEA show different amplitudes, the cortisol/DHEA molar ratio (i.e. the relative amounts of cortisol and DHEA in the saliva, and hence the brain) also varies during the day (Table 3.1). In the morning, the ratio is about 10, whereas in the evening it falls to 3–4. If this applies to the CSF, which seems very likely, this means that the brain is exposed to relatively much more cortisol than DHEA in the morning than the evening. This may have pathological significance under some circumstances (see below), if DHEA moderates the action of cortisol (as it seems to do).

Both cortisol and DHEA react to other states, such as illness or stress (see below). It is thus essential that the social and physiological circumstances of the subjects, as well as the time of day at which samples are collected, be specified. This is particularly important if the relations between cortisol and DHEA are a focus of study, since stress (e.g. illness) may increase cortisol but decrease DHEA (see below). Studies on hospital in-patients are thus likely to give information which may not be directly applicable to those in the community.

A single sample of either blood, saliva or CSF, however well characterized and documented, gives limited information. A series, taken at the same time (preferably on consecutive days) not only allows a more accurate estimate of mean values (reducing error due to technical and biological variability), but also allows a second parameter of value to be assessed. Day-to-day variation at a given time point, whether measured as "peaks" exceeding a given threshold value (e.g. 80th percentile of the overall mean) or a variance (if the number of samples is sufficient) indicate the stability or otherwise of hormone levels, and the frequency that they rise to levels that may (in the case of cortisol) occupy additional receptors. Clearly, it is possible to have mean values in two individuals that are similar, but the daily variation in one may be much greater than the other. This is likely to be biologically significant, given the relationship between CSF (brain) levels of steroid and receptor occupancy. This technique has been used to assess variations in cortisol in the saliva in a few studies (Gunnar et al., 1997), but DHEA in only one so far (Goodyer et al., 2000).

How many samples are needed to provide adequate data is still uncertain; we have used 4 daily samples for cortisol (and DHEA); but others report more (10 or more for salivary cortisol (Gunnar *et al.*, 1997). Further work is needed.

We now consider both experimental and clinical evidence for the neural actions of DHEA/S.

DHEA: EXPERIMENTAL EFFECTS ON BEHAVIOR

Aggression

DHEA/S has been shown to reduce male aggressiveness. In these behavioral studies, introducing lactating female mice to a cage of three resident castrated male mice results in the males attacking the female intruder. Subcutaneous administration of DHEA (80 µg/day for 15 days) reduced this aggressive behavior of the males. This effect does not seem to be due to conversion of DHEA to testosterone or to estrogenic metabolites (e.g. androst-5-en-3β, 17β diol). Moreover, repeating the experiments with the DHEA analog 3β-methyl-androst-5-en-17-one (M-DHEA), which cannot be metabolized into sex steroids, showed that M-DHEA was also able to inhibit aggressive behavior, at least as effectively as DHEA itself (Schlegel *et al.*, 1985; Haug *et al.*, 1989).

Cognition and learning

Several experimental reports have been published demonstrating memory-enhancing effects of DHEA and DHEAS. Briefly, these behavioral experiments on rats consisted of an active avoidance behavioral paradigm using buzzer foot shock in a T-maze. One week after training, retention for T-maze training was then tested. Intracerebroventricular (i.c.v.) injections of DHEA or DHEAS had enhancing effects on long-term retention of the T-maze training. The findings suggest that both DHEA and DHEAS can enhance learning and memory and alleviate amnesia by their actions on the CNS, either directly or via their metabolites (Roberts *et al.*, 1987; Flood *et al.*, 1988, 1992; Robel and Baulieu, 1995). However, in another study, chronic exposure to high levels of DHEAS (converted to DHEA *in vivo*) impaired long-term contextual fear conditioning, but did not affect short-term or auditory cue fear conditioning. Thus, in contrast to the behavioral tests described previously, DHEAS may have amnesiac properties for this type of learning. Therefore it seems DHEAS may have highly selective effects on behavior in different learning paradigms and memory tasks (Fleshner *et al.*, 1997). These authors also suggest that DHEAS exerted its effect by acting as a functional antiglucocorticoid in the processes that mediate learning and memory, because its effects on fear conditioning resembled those produced by adrenalectomy, and because the glucocorticoids receptor antagonists RU-38486 and RU-40555 also impaired contextual fear conditioning in the same way as DHEAS. The idea that DHEA is an antiglucocorticoid is an attractive one, and is discussed further in this chapter. This mechanism may explain many of the known neural actions of DHEA, and even the consequences of its age-related decline.

DHEA AND CORTISOL

There are other lines of experimental evidence suggesting that DHEA/S can act as a potent anti-glucocorticoid. For example, it antagonizes the immunosuppressant and lympholytic actions of cortisol (May *et al.*, 1990; Blauer *et al.*, 1991). Thus, circumstances in which DHEA/S is lowered (e.g. during illness or as part of the ageing process), may result in relative increases in unopposed cortisol, and hence the neurotoxic effects described below.

Before we discuss the role of DHEA in brain damage and protection, it is therefore necessary to consider, if briefly, the role of cortisol in these events, since much of the action of DHEA may be ascribed to its moderating the effects of glucocorticoids. Chronic corticosterone (the rodent equivalent to cortisol in man) exposure is neurotoxic, but the acute administration of even highly elevated levels of corticosterone alone does not kill neurons (Sapolsky, 1986; Behl, 1998). However, it can severely compromise the ability of the neurons to survive concurrent insults, a phenomenon termed "neuroendangerment" (Sapolsky, 1996; Gubba *et al.*, 2000), either by decreasing the threshold for toxicity or accelerating the damaging processes before survival mechanisms have a chance to operate. Most of the studies in this area concern glucocorticoid endangerment of the rat hippocampus, as this is the area of the brain richest in receptors for these steroids (McEwen *et al.*, 1968), although deleterious effects are not confined to this part of the brain. Glucocorticoid pre-treatment can potentiate powerfully the damaging effects of excitotoxins. The effects of kainic acid on induced *status epilepticus* and preferential CA3 hippocampal damage in rats were exacerbated by pre-exposure to either stress or elevated corticosterone levels (Stein and Sapolsky, 1988; Stein-Behrens *et al.*, 1992). Glucocorticoids also worsen quinolinic acid- and NMDA-induced lesions in the striatum (Supko and Johnston, 1994; Uhler *et al.*, 1994). These findings have been replicated *in vitro* on both primary hippocampal cultures (Packan and Sapolsky 1990; Sapolsky *et al.*, 1988) and hippocampal cell lines (Behl *et al.*, 1997), demonstrating that glucocorticoids can endanger neurons directly. Other insults that are accentuated by glucocorticoids include hypoglycaemia and metabolic toxins (Sapolsky, 1985, 1986; Sapolsky *et al.*, 1985), oxygen radical generators (McIntosh and Sapolsky, 1996; Goodman *et al.*, 1996; Behl *et al.*, 1997), and cholinergic or serotonergic toxins (Johnson *et al.*, 1989; Hortnagl *et al.*, 1993; Amoroso *et al.*, 1994). There may be a U-shaped effect of glucocorticoids on susceptibilty to brain damage: either very low or very high levels may both promote damage (Abraham *et al.*, 2000).

Neurogenesis in the hippocampus

The hippocampus has one remarkable feature, which may well have a bearing on its vulnerability to damage and its sensitivity to steroids. Neurogenesis persists in the dentate gyrus and in the ventricular subependymal zone well into adult life. First described many years ago (Altman and Das 1965), it is now apparent that the genesis of both neurons and glia continues to occur in adulthood, not only in rats but also in monkeys and humans (Eriksson *et al.*, 1998; Kornack and Rakic, 1999). This process seems to be modulated both by NMDA receptors and glucocorticoids

(Gould *et al.*, 1991; Cameron and McKay, 1999). Both stress and age impair hippo-campal neurogenesis (Fuchs and Flugge, 1998), whilst adrenalectomy stimulates it (Montaron *et al.*, 1999). Whether newly-formed neurons actually become part of the adult hippocampus, and what contribution they make to its function, is still unclear. Neurogenesis offers an important avenue through which stress, the ageing process and adrenal steroids can influence part of the brain known to be important for certain forms of memory.

DHEA and brain damage

There is accumulating evidence that neurosteroids such as DHEA may themselves contribute to brain damage or to the resilience of the brain to such damage. Preg-nenolone may exacerbate NMDA-induced degeneration of hippocampal neurons (Weaver Jr. *et al.*, 1998), but DHEA (and its sulfate, DHEAS) seem to be neuro-protective. DHEA has been shown to protect hippocampal neurons from excitotoxic cell death induced by either N-methyl-D-aspartic acid (NMDA), AMPA or Kainate *in vitro* (Kimonides *et al.*, 1998). In these studies, pre-treatment of primary hippocam-pal cultures with DHEA (10–100 nM) protected against NMDA-induced neuro-toxicity. DHEAS was also protective, though a 10-fold greater dose was required to be at least as effective as DHEA. In addition, *in vivo* studies on rats, showed that administration of DHEA, by implantation of subcutaneous pellets (which achieved levels equivalent to those found in young humans) protected the CA1/2 cells of the hippocampus against an unilateral lesion induced by either 5 or 10 nmol of NMDA (Figure 3.2). However, controversy remains about these neuroprotective effects of DHEA. Cardounel *et al.* (1999) have independently shown that 5 μM of DHEA was protective to cells of the clonal mouse hippocampal cell line HT-22 against

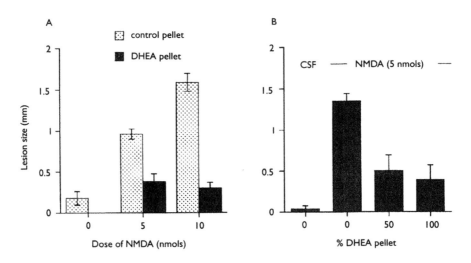

Figure 3.2 The effects of DHEA on local NMDA-induced lesions in the rat hippocampus. (A) The effects of DHEA on two doses of NMDA. (B) Dose-dependent effect of DHEA on NMDA-induced toxicity (from Kimonides *et al.*, 1998).

glutamate- and amyloid β-protein-induced toxicity. Bastianetto *et al.* (1999) also demonstrated that pre-treatment of rat primary mixed hippocampal cell cultures with DHEA (10–100 μM) protected against cell death induced by hydrogen peroxide and nitroprusside, thus demonstrating that in addition its anti-excito-toxic properties, DHEA has anti-oxidant actions as well. However the doses used in both these studies were 10^3 times greater than those used in the original Kimonides study. Indeed, in a subsequent follow-up study, Kimonides *et al.* (1999) showed that hippocampal cultures exposed to 500 nM of DHEA alone was found to significantly compromise neuronal survival. The reasons for this contradiction are unclear, but it may be that the level of toxicity induced in the Bastianetto study may have required higher doses of DHEA to be effective, and also the use of mixed hippocampal cultures could mean that glia may moderate the potency of DHEA. Nevertheless the supposition that DHEA/S is indeed neuroprotective is supported by the findings of Roberts *et al.* (1987) who showed that both DHEA (1 μM) and DHEAS (100 nM) enhanced neuronal and glial survival and differenti-ation in dissociated cultures of day 14 mouse embryo brain. Further contradictions to this complicated story have been published demonstrating that it is DHEAS, and not DHEA that is significantly protective against glutamate-induced toxicity *in vitro* in the hippocampus (Mao and Barger, 1998). In this study, however, the cultures were established from a mix of neocortical and hippocampal tissue, whereas in the Kimonides study neurons were isolated from dissected hippo-campi. The toxins and doses used also differ, with Mao and Barger exposing their cultures with 200 μM of glutamate, which required higher concentrations of DHEAS (3–300 nM) to provide protection. The experiments were also performed on younger cultures of 7–9 days, in comparison to those used in the original study (older than 10 days). It may be these technical differences explain these contrasting results.

DHEA/S also have demonstrated neurotrophic actions upon the brain. Compagnone and Mellon (1998) have shown that in primary cultures of mouse embryonic neocortical neurons, DHEA dose-dependently increased the length of neuritis containing the axonal marker Tau-1, and the number of variscosities and basket-like process formations, whereas DHEAS increased the length of neurites containing the dendritic marker microtubule-associated protein-2. Thus DHEA/S may have an additional role in the brain as signaling molecule for guiding cortical projections to their relevant targets. Whether DHEA is the active agent, or whether locally produced metabolies such as 7α-OH-DHEA (Akwa *et al.*, 1992), will prove to be more potent, awaits further work.

Although the neuroprotective ability of DHEA/S remains inconclusive, enough is known to suggest that lowered levels of DHEA or DHEAS due to age, stress or illness might cause increased vulnerability of the brain to neurological insults involving the excessive release of glutamate, including ischemia and other forms of neurodegeneration. This is one line of evidence supporting the proposition that supplementation with DHEA or DHEAS (normal constituents of the blood) might be effective in retarding or reversing neurological decline associated with ageing and illness.

Attempts to show that the possible protective effects of DHEA are more wide-spread in the brain, such as in the striatum, have so far proved unsuccessful.

DHEA *in vivo*, was unable to significantly protect against a unilateral striatal lesion induced by quinolinic acid, another glutamate analog (Ngai *et al.*, unpublished data). This suggests that the protective actions of DHEA may only be specific to the hippocampus. One explanation for this disparity could be that the protective effects of DHEA are only relevant to degeneration of certain brain areas or cell types. The hippocampus is a highly vulnerable brain region, and is much more susceptible to injury from excitotoxicity and glucocorticoid-induced toxicity compared to the striatum. The striatum, in contrast to the hippocampus is comparably resistant to injury, with a larger redundant cell population that may able to compensate for injury or cell death. However, more work in the striatum is needed (such as the use of striatal cell cultures), to specifically establish whether DHEA can only be effective in certain parts of the brain such as the hippocampus.

It must be recalled that the adult rodent, in which much of this work has been done, does not secrete appreciable amounts of DHEA. Thus one could argue that all the experiments carried out are non-physiological. As we have seen, there is a local source of DHEA in the rodent brain that is synthesized locally (Baulieu and Robel, 1990; Akwa *et al.*, 1991), though even this is disputed for the adult rodent brain. It remains to be determined whether these small concentrations produced in the rat brain (0.24 ± 0.33 ng/g) are indeed reliable and also functional.

DHEA as an antiglucocorticoid in the brain

There is evidence that DHEA does have anti-glucocorticoid actions in the brain, and can protect hippocampal neurons from glucocorticoid-induced toxicity (Kimonides *et al.*, 1999). In this study, DHEA (20–100 nM) was found to prevent neuronal cell death resulting from exposure of primary rat hippocampal cultures to corticosterone (20–500 nM) (Figure 3.3). The mechanism for this observed protective effect against glucocorticoid-induced is still not fully established, but it may be related to a set of enzymes called stress-activated protein kinases (SAPKs). These factors are activated by cellular stresses including those involving exposure to toxins, whereby they become phosphorylated and subsequently translocate to the nucleus to regulate the activity of transcription factors. One such factor that may be under the transcriptional control of SAPK3 is *c-jun*, which might be important in the sequence of intracellular pathways leading to either neuronal death or survival. DHEA was found to prevent glucocorticoid-induced nuclear translocation of SAPK3, suggesting that this may be important for glucocorticoid toxicity and the protective effect of DHEA. However, there is no direct evidence to prove that DHEA provides protection solely through this effect, and the physiological significance of glucocorticoid-induced nuclear translocation (regarding cell death or survival) is not yet known.

The mechanism for these anti-glucocorticoid actions is still unknown. Neither DHEA nor DHEAS were found to modulate glucocorticoid binding to receptors, however it has been tentatively shown that DHEA can reduce hepatic glucocorticoid receptors in the rat by 50% over 5 days of DHEA administration (Regelson *et al.*, 1988). There are however other hypotheses regarding the protective actions of DHEA and its role as an anti-glucocorticoid. Cardounel *et al.* (1999) have shown that treatment of cells of the mouse hippocampal cell line HT-22 with 5 µM

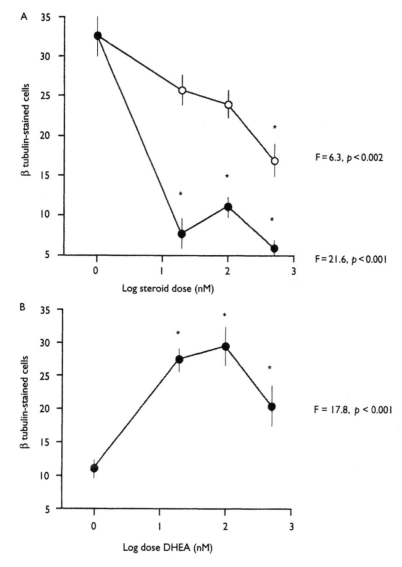

Figure 3.3 Modulation of the neurotoxic effects of corticosterone by DHEA on neurons in the cultures of rat hippocampal cells. (A) Dose-dependent effects of either DHEA (●) or corticosterone (○). (B) Prevention of the neurotoxic effects of 100 nM corticoster-one by increasing amounts of DHEA (from Kimonides *et al.*, 1999).

glutamate significantly increased glucocorticoid receptor (GR) nuclear localization in neurons, and this effect was reversed with pre-treatment of 5 μM DHEA for 24 hours before exposure to 5 mM glutamate. This was suggested to be in part the mechanism by which DHEA could protect hippocampal neurons from glutamate-induced toxicity. Thus it may be that DHEA can exert genomic effects that include changes in receptor gene transcription and expression in the brain, however, more work is needed to fully establish whether this truly is the case. DHEAS has also been shown to depress voltage-gated Ca^{2+} currents in adult mammalian hippocampal neurons (ffrench-Mullen and Spence, 1991), a process critical for mediating excitotoxic cell death (Coyle and Puttfarcken, 1993; Leist and Nicotera, 1998). In contrast, glucocorticoids have been shown to accentuate Ca^{2+} entry via voltage-gated channels (Elliott and Sapolsky, 1992). Also in the periphery, DHEA has been recently shown to enhance glucose uptake via activation of protein kinase C and phosphatidylinsortol-3 kinase (Ishizuka *et al.*, 1999), whereas it is well known that glucocorticoids inhibit glucose uptake, which is a key factor in its neurotoxic actions in the brain (Sapolsky, 1996). In humans, there is a marked increase in the cortisol/DHEA ratio with age, largely (but perhaps not exclusively) as the result of declining DHEA levels (Figure 3.4).

DHEA AND BRAIN FUNCTION IN MAN

The effects of DHEA on brain function in man can only be assessed indirectly, usually by studying some aspect of behavior. In some cases, this may give an indication of underlying brain disorder (e.g. the presence of Alzheimer's disease); in others, brain dysfunction has to be inferred (e.g. affective disorder). The interaction between DHEA and behavior is, like all hormones, two-directional: changes in behavior may alter DHEA (i.e. as a state-dependent factor); or changes in DHEA (either as the result of some underlying process such as ageing, or as a secondary consequence of state-induced alterations) may have an effect on brain function. The only reliable way to establish a causative link between DHEA/S levels and either the presence or onset of brain dysfunction (however assessed) is to administer DHEA/S to subjects as part of a well-controlled and relevant trial with adequate duration and sensitive end-points.

Age-related decline in DHEA in man: correlation with behavior

The fact that both DHEA and a variety of cognitive and other neural functions decline with age is not, of course, evidence for an association between them: only a suggestion that this might be the case. More pertinent is evidence that individual differences in the age-related decline in DHEA or individual differences in levels at some predetermined time point either predict or are associated with equivalent differences in the onset or incidence of some dependent disorder. A number of studies suggest this to be the case. A large survey of males aged 39–70 showed that chronic illness was associated with lower DHEAS, though the direction of this relation cannot be determined from this data (Gray *et al.*, 1991); further analysis showed that lower DHEAS was associated with age-related impotence (Feldman *et al.*, 1994). In a mixed-sex group of subjects over 65 years: DHEAS levels were negatively related over four years to a measure of "well-being", negative mood states and mortality rates (Berr *et al.*, 1996). Other ageing parameters, not immediately relevant to brain function, such as cardiovascular disease and type 2 (maturity-onset) diabetes, have also been related to relatively lower levels of DHEA/S.

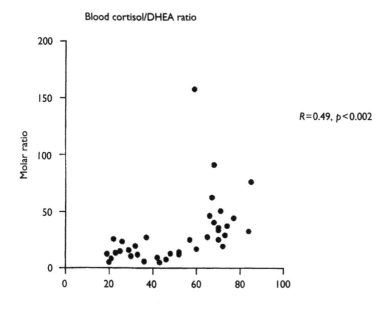

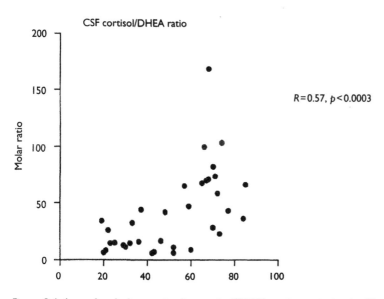

Figure 3.4 Age-related changes in the cortisol/DHEA molar ratio in the blood and CSF of humans (from Guazzo *et al.*, 1996).

DHEA and stress

Stressful events are a major source of environmental stimuli altering adrenal steroids. Blood levels of cortisol and DHEA/S seem to be altered in opposite directions by stress. Cortisol is increased by stress; that is, physiological or behavioral demand outside the ordinary (e.g. physical or social adversity, and illness, including infection, trauma and mental disorder such as major depression) (Sachar, 1976; Selye, 1978; Herbert, 1987; Munck *et al.*, 1984). In contrast, DHEA/S levels may be reduced by acute or persistent stress, or somatic and mental illness – for example following admission to an acute cardiac intensive care unit (Spratt *et al.*, 1993), in patients with rheumatoid arthritis (Hedman *et al.*, 1992; Hall *et al.*, 1997), in army officers under severe training (this was prevented by stress-reduction therapy) (Littman *et al.*, 1993). DHEA levels decreased, but DHEAS increased, in young military cadets during a relatively brief stress (5 days arduous training) (Opstad, 1992), though this included food deprivation. 14 weeks of exercise training did not alter DHEAS levels in middle-aged men (Houmard *et al.*, 1994), suggesting that physical exertion *per se* is not an effective DHEA/S suppressant. Not everyone finds decreased DHEA after stress: for example, a study of acute stress (parachute jumps) showed increased blood DHEA, though this study was unusual in failing to find increased blood cortisol (Oberbeck *et al.*, 1998).

DHEA, cognition and neurodegenerative conditions

The fact that DHEA may be neuroprotective suggests that there may be positive associations with age-related decrease in DHEA and likelihood of cognitive decline. In fact, the evidence that this is so is mixed: a recent study on older men showed that those with relatively lower salivary DHEA had lower scores in cognitive tests for verbal memory (van Neikerk *et al.*, 2001). However, attempts to relate lower DHEA/S levels with the presence of Alzheimer's disease (AD) have been inconclusive: whereas an early report found decreased DHEAS levels in AD (Sunderland, 1989), this has not been confirmed by more recent ones (Leblhuber *et al.*, 1990; Schneider *et al.*, 1992; Carlson *et al.*, 1999), though the latter did find that AD patients with higher DHEAS levels performed somewhat better on a memory test (name recall). It should be evident that, even if there is no difference between DHEA/S levels in those with AD and appropriate controls, this in no way excludes age-related decline in DHEA/S as a precipitating or risk factor for developing AD in vulnerable people. An adequate trial of long-term DHEA supplementation in such subjects prior to the onset of AD – the only way to establish whether there is a causative link between declining DHEA(S) and dementia – has not been reported.

However, so far, giving supplementary DHEA to ageing humans has not improved cognition (reviewed by Huppert *et al.* (2000). Some of these studies are rather too short (e.g. 2 weeks) to represent a reasonable test of this hypothesis (Kudielka *et al.*, 1998; Wolf *et al.*, 1998), but a more recent one in which older men (62–76 years) were given 50 mg/day DHEA for 13 weeks also failed to show changes in a range of cognitive measures (including visuo-spatial memory, speed of response and episodic memory) (van Niekerk *et al.*, 2001). Again, it must be emphasised that these negative results (which need amplification using longer treatment periods and even more refined methods of analysis) do not exclude age-dependent reductions in DHEA/S as a precipitating or permissive factor in the cognitive decline typical of the ageing brain. Only prospective trials of DHEA supplementation of adequate power can establish whether this is a significant property of the ageing process.

DHEA and mood

A variety of studies have shown a negative relation between DHEA and low affective state (either well-being or depression), though there are exceptions. The results of large community surveys have already been discussed: they indicate such a relation. In older men (62–76 years) a higher morning cortisol and a morning cortisol/DHEA ratio was associated with greater general mood disturbance, confusion and higher evening DHEA was associated with less anxiety (van Niekerk *et al.*, 2001). In both adults and adolescents with unipolar major depressive disorder (MDD), lower morning (8.00 a.m.) levels of DHEA in the saliva have been found: this is not the consequence of concurrent drug treatment (Figure 3.5) (Goodyer *et al.*, 1996; Michael *et al.*, 2000). There have, however, been opposing findings: higher levels of DHEAS and evening levels of DHEA have been reported in MDD (Heuser *et al.*, 1998; Takebayashi *et al.*, 1998). In adolescents, a higher cortisol/DHEA ratio is found during illness (Figure 3.5) and is associated with delayed recovery (Herbert *et al.*, 1996). Giving DHEA may also enhance mood. Three months DHEA supplementation (50 mg/day) in middle-aged subjects (aged 40–70 years) improved "well-being" (Morales *et al.*, 1994), though, as has been repeatedly pointed out, the methods of measuring this index and its definition are not clear. However, subsequent studies have tended to confirm it. Twelve months treatment (50 mg/day) given to men and women aged 60–79 years improved well-being and sexual interest (Baulieu *et al.*, 1999). A small double-blind study of DHEA supplementation (up to 90 mg/day) in 11 patients with MDD resulted in decrease depressive symptomatology (Wolkowitz *et al.*, 1999). Addison's disease is customarily treated with cortisol and a mineralocorticoid, but not DHEA: such patients, therefore, are almost totally deficient in this steroid and offer a convenient model for testing the function of DHEA/S in man. Two studies have now shown that adding DHEA to the treatment regime of such patients for a few months improves mood, "well-being", sexual function and fatigue (Arlt *et al.*, 1999; Hunt *et al.*, 2000). This is the clearest demonstration so far that DHEA/S has a physiological role in man, and thus that age-related decreases in DHEA/S levels may have behavioral and neurobiological significance for the alterations seen in advancing years. It may also indicate that more transient reductions in DHEA, as during stress or depressive illness, may also be neurologically significant. How far these changes depend upon, or relate to, the actions of cortisol remain to be fully defined, though this appears to be the most attractive line of enquiry at the moment. It also seems, at the moment, that the relation between DHEA and mood is more secure than that for cognitive functions though this conclusion may well be modified in the light of more searching investigations.

CONCLUSION

Is the current information on DHEA persuasive that it has a role to play in age-related neural decline? The experimental evidence lacks several critical features. There is no robust and repeatable model for demonstrating that DHEA protects the brain against either "natural" (age-related) or experimentally induced damage. Rodents have little DHEA, so that replacement in these species will always be

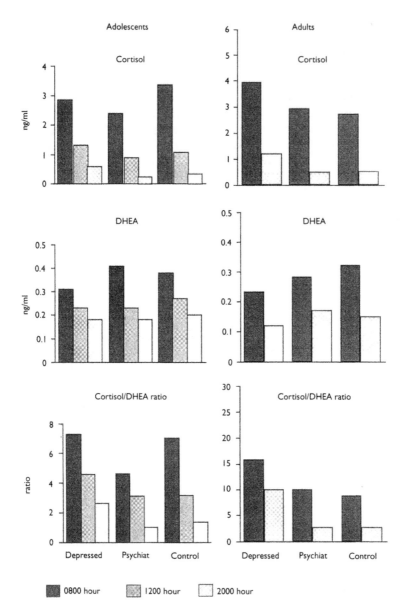

Figure 3.5 Alteration in salivary cortisol and DHEA, and the resultant cortisol/DHEA ratios, in patients with major depressive disorder (MDD). MDD is associated with lower morning DHEA, higher evening cortisol and increased cortisol/DHEA ratios. (Left panel) Results at three time points from drug-free depressed adolescents, compared with either those with other psychiatric conditions, or controls. (from Goodyer, *et al.*, 1996). (Right panel) Results at two time points from adult MDD patients receiving anti-depressants, compared with recovered patients still receiving drug therapy or controls (from Michael *et al.*, 2000).

open to the objection that it is not physiological. The relative role of peripheral and central "neurosteroids" (including DHEA) is still debatable. Although there is evidence that DHEA acts both directly on the $GABA_A$ receptor, and indirectly on the brain via modulating glucocorticoids, the mechanisms for either action are still obscure.

The clinical data lacks satisfactory long-term measures of replacement DHEA, to determine whether this might deter "normal" age-related decline, or prevent or reduce it in those especially vulnerable to cognitive failure (e.g. Alzheimer's disease). Without an adequate trial (sufficient in number, design, duration and endpoints) this issue will never be settled. We have repeatedly pointed out in this chapter that there is enough preliminary evidence to warrant such a study, but no substitute for it. Nevertheless, it is our opinion that the possibility remains that – subject to further evidence – DHEA replacement therapy, at least in a vulnerable section of ageing people of both sexes, may become, in time, as acceptable as HRT therapy in post-menopausal women. This may open a new chapter in our understanding of the ageing process in the brain and hopefully lead to the alleviation of some of the problems associated with declining neural function with age.

ACKNOWLEDGMENTS

The work of our laboratory is supported by grants from the Wellcome Trust, BBSRC and MRC. Lai-Yeung Ngai is supported by an MRC Research Studentship.

REFERENCES

Abraham, I., Harkany, T., Horvath, K.M., Veenema, A.H., Penke, B., Nyakas, C. *et al.* (2000) Chronic corticosterone administration dose-dependently modulates Ab(1–42)- and NMDA-induced neurodegeneration in rat magnocellular nucleus basalis. *J. Neuroendocrinol.*, **12**, 486–494.

Akwa, Y., Morfin, R.F., Robel, P. and Baulieu, E.-E. (1992) Neurosteroid metabolism: 7α-hydroxylation of dehydroepiandrosterone and pregnenolone by rat brain microsomes. *Biochem. J.*, **288**, 959–964.

Akwa, Y., Young, J., Kabbadi, K., Sancho, M.J., Zucman, D., Vourc'h, C. *et al.* (1991) Neurosteroids: Biosynthesis, metabolism and function of pregnenolone and dehydroepiandrosterone in the brain. *J. Steroid Biochem.*, **40**, 71–81.

Altman, J. and Das, G.D. (1965) Autoradiographic and histological evidence of postnatal hippocampal neurogenesis in rats. *J. Comp. Neurol.*, **124**, 319–335.

Amoroso, D., Kindel, G., Wulfert, E. and Hanin, I. (1994) Long-term exposure to high levels of corticosterone aggravates AF64A-induced cholinergic hypofunction in rat hippocampus *in vivo*. *Brain Res.*, **657**, 227–235.

Arlt, W., Callies, F., van Vlijmen, J.C., Koehler, I., Reicke, M., Bidlingmaier, M. *et al.* (1999) Dehydroepiandrosterone replacement in women with adrenal insufficiency. *New Engl. J. Med.*, **341**, 1013–1020.

Bastianetto, S., Ramassamy, C., Poirier, J. and Quirion, R. (1999) Dehydroepiandrosterone (DHEA) protects hippocampal cells from oxidative stress-induced damage. *Brain Res. Mol. Brain Res.*, **66**, 35–41.

Baulieu, E.-E. (1998) Neurosteroids: a novel function of the brain. *Psychoneuroendocrinology*, **23**, 963–987.

Baulieu, E.-E. and Robel, P. (1990) Neurosteroids: A new brain function? *J. Steroid Biochem.*, **37**, 395–403.

Baulieu, E.-E., Thomas, G., Legrain, S., Lahlou, N., Roger, M., Debuire, B. *et al.* (1999) Dehydroepiandrosterone (DHEA), DHEA sulfate, and aging: contribution of the DHEAge study to a sociomedical issue. *Proc. Natl. Acad. Sci. USA*, **97**, 4279–4284.

Behl, C. (1998) Effects of glucocorticoids on oxidative stress-induced hippocampal cell death: implications for the patheogenesis of Alzheimer's disease. *Exp. Gerontol.*, **33**, 689–696.

Behl, C., Lezoualc'h, F., Trapp, T., Widmann, M., Skutella, T. and Holsboer, F. (1997) Glucocorticoids enhance oxidative stress-induced death in hippocampal neurons *in vitro*. *Endocrinology*, **138**, 101–106.

Berr, C., Lafont, S., Debuire, B., Dartigues, J.F. and Baulieu, E.-E. (1996) Relationships of dehydroepiandrosterone sulfate in the elderly with functional, psychological, and mental status, and short-term mortality: a French community-based study. *Proc. Natl. Acad. Sci. USA*, **93**, 13410–13415.

Blauer, K.L., Poth, M., Rogers, W. and Bernton, E.W. (1991) Dehydroepiandrosterone antagonizes the suppressive effects of dexamethasone on lymphocyte proliferation. *Endocrinology*, **129**, 3174–3179.

Cameron, H.A. and McKay, R.D.G. (1999) Restoring production of hippocampal neurons in old age. *Nature Neurosci.*, **2**, 894–897.

Cardounel, A., Regelson, W. and Kalimi, M. (1999) Dehydroepiandrosterone protects hippocampal neurons against neurotoxin-induced cell death: mechanism of action. *Proc. Soc. Exp. Biol. Med.*, **222**, 145–149.

Carlson, L.E., Sherwin, B.B. and Chertkow, H.M. (1999) Relationships between dehydro-epiandrosterone sulfate (DHEAS) and cortisol (CRT) plasma levels and everyday memory in Alzheimer's disease patients compared to healthy controls. *Horm. Behav.*, **35**, 254–263.

Compagnone, N.A. and Mellon, S.H. (1998) Dehydroepiandrosterone: a potenetial signalling molecule for neocortical organisation during development. *Proc. Natl. Aacad. Sci. USA*, **95**, 4678–4683.

Corpéchot, C., Robel, P., Axelson, M., Sjövall, J. and Baulieu, E.-E. (1981) Characterization and measurement of dehydroepiandrosterone sulfate in rat brain. *Proc. Natl. Acad. Sci. USA*, **78**, 4704–4707.

Corpéchot, C., Synguelakis, M., Talha, S., Axelson, M., Sjövall, J., Vihko, R. *et al.* (1983) Pregnenolone and its sulfate ester in the rat brain. *Brain Res.*, **270**, 119–125.

Coyle, J.T. and Puttfarcken, P. (1993) Oxidative stress, glutamate, and neurodegenerative disorders. *Science*, **262**, 689–695.

del Campo, A.F.M., Dowson, J.H., Herbert, J. and Paykel, E.S. (2000) Diurnal variations in endocrine and psychological responses to 0.2 mg/kg naloxone administration in patients with major depressive disorder and matched controls. *J. Affect. Disorders*, **57**, 37–47.

Diczfalusy, E. (1984) The early history of estriol. *J. Steroid Biochem.*, **20**, 945–953.

Elliott, E.M. and Sapolsky, R.M. (1992) Corticosterone enhances kainic acid-induced calcium elevation in cultures hippocampal neurons. *J. Neurochem.*, **59**, 1033–1040.

Eriksson, P.S., Perfilieva, E., Bjork-Eriksson, T., Alborn, A.M., Nordborg, C., Peterson, D.A. *et al.* (1998) Neurogenesis in the adult human hippocampus. *Nat. Med.*, **4**, 1313–1317.

Feldman, H.A., Goldstein, I., Hatzichristou, D.G., Krane, R.J. and McKinlay, J.B. (1994) Impotence and its medical and psychosocial correlates: results of the Massachusetts male aging study. *J. Urology*, **151**, 54–61.

ffrench-Mullen, J.M.H. and Spence, K.T. (1991) Neurosteroids block Ca^{2+} channel current in freshly isolated hippocampal CA1 neurons. *Eur. J. Pharmacol.*, **202**, 269–272.

Fleshner, M., Pugh, C.R., Tremblay, D. and Rudy, J.W. (1997) DHEA-S selectively impairs contextual-fear conditioning: support for the antiglucocorticoid hypothesis. *Behav. Neurosci.*, **111**, 512–517.

Flood, J.F., Morley, J.E. and Roberts, E. (1992) Memory-enhancing effects in male mice of pregnenolone and steroids metabolically derived from it. *Proc. Natl. Acad. Sci. USA*, **89**, 1567–1571.

Flood, J.F., Smith, G.E. and Roberts, E. (1988): Dehydroepiandrosterone and its sulfate enhance memory retention in mice. *Brain Res.*, **447**, 269–278.

Fuchs, E. and Flugge, G. (1998) Stress, glucocorticoids and structural plasticity of the hippocampus. *Biobehav. Rev.*, **23**, 295–300.

Goodman, Y., Bruce, A.J., Cheng, B. and Mattson, M.P. (1996) Estrogens attentuate and corticosterone exacerbates excitotoxicity, oxidative injury, and amyloid β-peptide toxicity in hippocampal neurons. *J. Neurochem.*, **66**, 1836–1844.

Goodyer, I.M., Herbert, J., Altham, P.M.E., Pearson, J., Secher, S.M. and Shiers, H.M. (1996) Adrenal secretion during major depression in 8- to 16-year-olds, I. Altered diurnal rhythms in salivary cortisol and dehydroepiandrosterone (DHEA) at presentation. *Psychol. Med.*, **26**, 245–256.

Goodyer, I.M., Herbert, J., Tamplin, A. and Altham, P.M.E. (2000) First-episode major depression in adolsecents. Affective, cognitive and endocrine characteristics of risk statis and predictors of onset. *Brit. J. Psychiat.*, **176**, 142–149.

Gould, E., Woolley, C.S., Cameron, H.A., Daniels, D.C. and McEwen, B.S. (1991) Adrenal steroids regulate postnatal development of the rat dentate gyrus: II. Effects of glucocorticoids and mineralocorticoids on cell birth. *J. Comp. Neurol.*, **313**, 486–493.

Gray, A., Feldman, H.A., McKinlay, J.B. and Longcope, C. (1991) Age, disease, and changing sex hormone levels in middle-aged men: results of the Massachusetts male aging study. *J. Clin. Endocr. Metab.*, **73**, 1016–1025.

Guazzo, E.P., Kirkpatrick, P.J., Goodyer, I.M., Shiers, H.M. and Herbert, J. (1996) Cortisol, dehydroepiandrosterone (DHEA), and DHEA sulfate in the cerebrospinal fluid of man: relation to blood levels and the effects of age. *J. Clin. Endocr. Metab.*, **81**, 3951–3960.

Gubba, E.M., Netherton, C.M. and Herbert, J. (2000) Endangerment of the brain by glucocorticoids: experimental and clinical evidence. *J. Neurocytol.*, **29**, 439–449.

Gunnar, M.R., Tout, K., de Haan, M., Pierce, S. and Stansbury, K. (1997) Temperament, social competence, and adrenocortical activity in preschoolers. *Dev. Psychobiol.*, **31**, 65–85.

Hall, F., Humby, T., Wilkinson, L. and Robbins, T. (1997) The effects of isolation-rearing on preference by rats for a novel environment. *Physiol. Behav.*, **61**, 299–303.

Haug, M., Ouss-Schlegel, M.L., Spetz, J.F., Brain, P.F., Simon, V., Baulieu, E.-E. *et al.* (1989) Suppressive effects of dehydroepiandrosterone and 3-beta-methylandrost-5-en-17-one on attack towards lactating female intruders by castrated male mice. *Physiol. Behav.*, **46**, 955–959.

Hedman, M., Nilsson, E. and Delatorre, B. (1992) Low blood and synovial-fluid level of sufoconjugated steroids in rheumatoid arthritis. *Clin. Exp. Rheum.*, **10**, 25–30.

Herbert, J. (1987) Neuroendocrine responses to social stress. In A. Grossman (ed), *Neuroendocrinology of Stress. Clinical Endocrinology and Metabolism*, **2**, 467–490.

Herbert, J., Goodyer, I.M., Altham, P.M., Pearson, J., Secher, S.M. and Shiers, H.M. (1996) Adrenal secretion and major depression in 8- to 16-year-olds, II. Influence of co-morbidity at presentation. *Psychol. Med.*, **26**, 257–263.

Heuser, I., Deuschle, M., Luppa, U., Schweiger, U., Standhardt, H. and Weber, B. (1998) Increased diurnal plasma concentrations of dehydroepiandrosterone in depressed patients. *J. Clin. Endocr. Metab.*, **83**, 3130–3133.

Hortnagl, H., Berger, M., Havelec, L. and Hornykiewicz, O. (1993) Role of glucocorticoids in the cholinergic degeneration in rat hippocampus induced by ethylcholine aziridinium (AF64A). *J. Neurosci.*, **13**, 2939–2946.

Houmard, J.A., McCulley, C., Shinebarger, M.H. and Bruno, N.J. (1994) Effects of exercise training on plasma androgens in men. *Horm. Metab. Res.*, **26**, 297–300.

Hunt, P.J., Gurnell, E.M., Huppert, F.A., Richards, C., Prevost, T., Wass, J.A., Herbert, J. and Chatterjee, V.K. (2000) Improvement in mood and fatigue following DHEA replacement in a randomised double-bind trial in Addison's disease. *J. Clin. Endocr. Metab.*, **85**, 4650–4656.

Huppert, F.A., Van Niekerk, J.K. and Herbert, J. (2000) Dehydroepiandrosterone (DHEA) supplementation for cognition and well-being. *Cochrane Database Systems Reviews*, **2**.

Ishizuka, T., Kajita, K., Miura, A., Ishizawa, M., Kanoh, Y., Itaya, S. *et al.* (1999) DHEA improves glucose uptake via activations of protein kinase C and phosphatidylinositol 3-kinase. *Am. J. Physiol.*, **276**, E196–E204.

Johnson, M., Stone, D., Bush, L., Hanson, G. and Gibb, J. (1989) Glucocorticoids and 3,4-methylenedioxymethamphetamine (MDMA)-induced neurotoxicity. *Eur. J. Pharmacol.*, **161**, 181.

Kimonides, V.G., Khatibi, N.H., Svendsen, C.N., Sofroniew, M.V. and Herbert, J. (1998) Dehydroepiandrosterone (DHEA) and DHEA-sulfate (DHEAS) protect hippocampal neurons against excitatory amino acid-induced neurotoxicity. *Proc. Natl. Acad. Sci. USA*, **95**, 1852–1857.

Kimonides, V.G., Spillantini, M.G., Sofroniew, M.V., Fawcett, J.W. and Herbert, J. (1999) Dehydroepiandrosterone antagonizes the neurotoxic effects of corticosterone and translocation of stress-activated protein kinase 3 in hippocampal primary cultures. *Neuroscience*, **89**, 429–436.

Kornack, D.R. and Rakic, P. (1999) Continuation of neurogenesis in the hippocampus of the adult macaque monkey. *Proc. Natl. Acad. Sci. USA*, **96**, 5768–5773.

Kudielka, B.M., Hellhammer, J., Hellhammer, D.H., Wolf, O.T., Pirke, K.M., Varadi, E. *et al.* (1998) Sex differences in endocrine and psychological responses to psychosocial stress in healthy elderly subjects and the impact of a 2-week dehydroepiandrosterone treatment. *J. Clin. Endocr. Metab.*, **83**, 1756–1761.

Leblhuber, F., Windhager, E., Reisecker, F., Steinparz, F.X. and Dienstl, E. (1990) Dehydroepiandrosterone sulphate in Alzheimer's disease. *Lancet*, **336**, 449–450.

Leist, M. and Nicotera, P. (1998) Apoptosis, excitotoxicity, and neuropathology. *Exp. Cell Res.*, **239**, 183–201.

Littman, A.B., Fava, M., Halperin, P., Lamon-Fava, S., Drews, F.R., Oleshansky, M.A. *et al.* (1993) Physiologic benefits of a stress reduction program for healthy middle-aged army officers. *J. Psychosom. Res.*, **37**, 345–354.

Mao, X. and Barger, S.W. (1998) Neuroprotection by dehydroepiandrosterone-sulfate: role of an NFkB like factor. *NeuroReport*, **9**, 759–763.

Martensz, N.D., Herbert, J. and Stacey, P.M. (1983) Factors regulating levels of cortisol in cerebrospinal fluid of monkeys during acute and chronic hypercortisolemia. *Neuroendocrinology*, **36**, 39–48.

May, M., Holmes, E., Rogers, W. and Poth, M. (1990) Protection from glucocorticoid induced thymic involution by dehydroepiandrosterone. *Life Sci.*, **46**, 1627–1631.

McEwen, B.S., Weiss, J.M. and Schwartz, L.S. (1968) Selective retention of corticosterone by limbic structures in rat brain. *Nature*, **220**: 911–912.

McIntosh, L.J. and Sapolsky, R.M. (1996) Glucocorticoids increase the accumulation of reactive oxygen species and enhance adriamycin-induced toxicity in neuronal culture. *Exp. Neurol.*, **141**, 201–206.

Mellon, S.H. and Deschepper, C.F. (1993) Neurosteroid biosynthesis: genes for adrenal steroidogenic enzymes are expressed in the brain. *Brain Res.*, **629**, 283–292.

Michael, A., Jenaway, A., Paykel, E.S. and Herbert, J. (2000) Altered salivary DHEA levels in major depression in adults. *Biol. Psychiat.*, **48**, 989–995.

Montaron, M.F., Petry, K.G., Rodriguez, J.J., Marinelli, M., Aurousseau, C., Rougon, G. *et al.* (1999) Adrenalectomy increase neurogenesis but not PSA-NCAM expression in aged dentate gyrus. *Eur. J. Neurosci.*, **11**, 1479–1485.

Morales, A.J., Nolan, J.J., Nelson, J.C. and Yen, S.S.C. (1994) Effects of replacement dose of dehydroepiandrosterone in men and women of advancing age. *J. Clin. Endocr. Metab.*, **78**, 1360–1367.

Munck, A., Guyre, P.M. and Holbrook, N.J. (1984) Physiological functions of glucocorticoids is stress and their relation to pharmacological actions. *Endocr. Rev.*, **5**, 25–44.

Oberbeck, R., Benschop, R.J., Jacobs, R., Hosch, W., Jetschmann, J.U., Schurmeyer, T.H. *et al.* (1998) Endocrine mechanisms of stress-induced DHEA-secretion. *J. Endocrinol. Invest.*, **21**, 148–153.

Opstad, P.K. (1992) The hypothalamo-pituitary regulation of androgen secretion in young men after prolonged physical stress combined with energy and sleep deprivation. *Acta Endocrinol. (Copenh.)*, **127**, 231–236.

Orentreich, N., Brind, J.L., Rizer, R.L. and Vogelman, J.H. (1984) Age changes and sex differences in serum dehydroepiandrosterone sulfate concentrations throughout adulthood. *J. Clin. Endocr. Metab.*, **59**, 551–555.

Orentreich, N., Brind, J.L., Vogelman, J.H., Andres, R. and Baldwin, H. (1992) Long-term longitudinal measurements of plasma dehydroepiandrosterone sulfate in normal men. *J. Clin. Endocr. Metab.*, **75**, 1002–1004.

Packan, D.R. and Sapolsky, R.M. (1990) Glucocorticoid endangerment of the hippocampus: tissue, steroid and receptor specificity. *Neuroendocrinology*, **51**, 613–618.

Parker, L.N. (1991) Adrenarche. *Endocr. Metab. Clin.*, **20**, 71–83.

Parker, L.N. (1993) Adrenarche. *The Endocrinologist*, **3**, 385–391.

Regelson, W., Loria, R. and Kalimi, M. (1988) Hormonal intervention: "buffer hormones" or "state dependency" the role of dehydroepiandrosterone (DHEA), thyroid hormone, estrogen and hypophysectomy in aging. *Ann. N. Y. Acad. Sci.*, **521**, 260–273.

Robel, P. and Baulieu, E.-E. (1994) Dehydroepiandrosterone (DHEA) is a neuroactive neurosteroid. *Ann. N. Y. Acad. Sci.*, **774**, 82–110.

Robel, P., Young, J., Corpechot, C., Mayo, W., Perche, F., Haug, M. *et al.* (1995) Biosyntheis and assay of neurosteroids in rats and mice: functional correlates. *J. Steroid Biochem.*, **53**, 335–360.

Robel, P. and Baulieu, E.-E. (1995) Neurosteroids: biosynthesis and function. *Crit. Rev. Neurobiol.*, **9**, 383–394.

Robel, P., Bourreau, E., Corpéchot, C., Dang, D.C., Halberg, F., Clarke, C. *et al.* (1987) Neuro-steroids: 3 beta-hydroxy-delta 5 derivatives in rat and monkey brain. *J. Steroid Biochem.*, **27**, 649–655.

Roberts, E., Bologa, L., Flood, J.F. and Smith, G.E. (1987) Effects of dehydroepiandrosterone and its sulfate on brain tissue in culture and on memory in mice. *Brain Res.*, **406**, 357–362.

Sachar, E.J. (1976) Neuroendocrine dysfunction in depressive illness. *Ann. Rev. Med.*, **27**, 389–396.

Sapolsky, R.M. (1985) A mechanism for glucocorticoid toxicity in the hippocampus: increased neuronal vulnerability to metabolic insults. *J. Neurosci.*, **5**, 1228–1232.

Sapolsky, R.M. (1986) Glucocorticoid toxicity in the hippocampus: reversal by supplementation with brain fuels. *J. Neurosci.*, **6**, 2240–2244.

Sapolsky, R.M. (1996) Stress, Glucocorticoids, and damage to the nervous system: the current state of confusion. *Stress*, **1**, 1–19.

Sapolsky, R.M., Krey, L.C. and McEwen, B.S. (1985) Prolonged glucocorticoid exposure reduces hippocampal neuron number: implications for aging. *J. Neurosci.*, **5**, 1222–1227.

Sapolsky, R.M., Packan, D.R. and Vale, W.W. (1988) Glucocorticoid toxicity in the hippocampus: *in vitro* demonstration. *Brain Res.*, **453**, 367–371.

Schlegel, M.L., Spetz, J.F., Robel, P. and Haug, M. (1985) Studies on the effects of dehydroepiandrosterone and its metabolites on attack by castrated mice on lactating intruders. *Physiol. Behav.*, **34**, 867–870.

Schneider, L.S., Hinsey, M. and Lyness, S. (1992) Plasma dehydroepiandrosterone sulfate in Alzheimer's disease. *Biol. Psychiat.*, **31**, 205–208.

Selye, H. (1978): *The stress of life*. New York: McGraw-Hill.

Spratt, D.I., Longcope, C., Cox, P.M., Bigos, S.T. and Wilbur-Welling, C. (1993) Differential changes in serum concentrations of androgens and estrogens (in relation with cortisol) in postmenopausal women with acute illness. *J. Clin. Endocr. Metab.*, **76**, 1542–1547.

Stein, B.A. and Sapolsky, R.M. (1988) Chemical adrenolectomy reduces hippocampal damage induced by kainic acid. *Brain Res.*, **473**, 175–180.

Stein-Behrens, B.A., Elliott, E.M., Miller, C.A., Schilling, J.W., Newcombe, R. and Sapolsky, R.M. (1992) Glucocorticoids exacerbate kainic acid-induced extracellular accumulation of excitatory amino acids in the rat hippocampus. *J. Neurochem.*, **58**, 1730–1735.

Sunderland, T., Merril, C.R., Harrington, M.G., Lawlor, B.A., Molchan, S.E., Martinez, R. *et al.* (1989) Reduced plasma dehydroepiandrosterone concentrations in Alzheimer's disease. *Lancet*, **2**, 570.

Supko, D.E. and Johnston, M.V. (1994) Dexamethasone potentiates NMDA receptor-mediated neuronal injury in the postnatal rat. *Eur. J. Pharmacol.*, **270**, 105–113.

Takebayashi, M., Kagaya, A., Uchitomi, Y., Kugaya, A., Puraoka, M., Yokota, N. *et al.* (1998) Plasma dehydroepiandrosterone sulfate in unipolar major depression. *J. Neurotransm.*, **105**, 537–542.

Uhler, T.A., Frim, D.M., Pakzaban, P. and Isacson, O. (1994) The effects of megadose methylprednisolone and U-78517F on toxicity mediated by glutamate receptors in the rat neostriatum. *Neurosurgery*, **34**, 122–128.

van Niekerk, J.K., Huppert, F.A. and Herbert, J. (2001) Salivary cortisol and DHEA: association with measures of cognition and well-being in normal oldermen, and effects of three months of DHEA supplementaion. *Psychoneuroendocrinology*, **26**, 591–612.

Weaver, Jr. C.E., Wu, F.S., Gibbs, T.T. and Farb, D.H. (1998) Pregnenolone sulfate exacerberates NMDA-induced death of hippocampal neurons. *Brain Res.*, **803**, 129–136.

Winter, J.S.D. (1998) Fetal and neonatal adrenocortical physiology. In R.A. Polin and W.W. Fox (eds), *Fetal and Neonatal Physiology*, 2nd edn. Baltimore: W.B. Saunders, pp. 2447–2459.

Wolf, O.T., Naumann, E., Hellhammer, D.H. and Kirschbaum, C. (1998) Effects of dehydroepiandrosterone replacement in elderly men on event-related potentials, memory, and well-being. *J. Gerontol.*, **53**, M385–M390.

Wolkowitz, O.M., Reus, V.I., Keebler, A., Nelson, N., Friedland, M., Brizendine, L. *et al.* (1999) Double-blind treatment of major depression with dehydroepiandrosterone. *Am. J. Psychiat.*, **156**, 646–649.

Zwain, I.H. and Yen, S.S. (1999) Neurosteroidogenesis in astrocytes, oligodendrcytes, and neurons of cerebral cortex of rat brain. *Endocrinology*, **140**, 3843–3852.

Chapter 4

DHEA: hormone of youth and resilience – still a maverick

Maria Dorota Majewska

INTRODUCTION: THE NEUROSTEROIDS

The exploding field of neurosteroid research, conducted both at basic and clinical levels, was created on the base of two lines of discoveries from the 1980s. The group of Corpechot, Baulieu and Robel reported the existence of steroidogenesis in the Central Nervous System (Corpechot *et al.*, 1981, 1983; Baulieu *et al.*, 1987) and created the term of "neurosteroids", and independently Harrison with Simmons and I discovered steroid interaction with neurotransmitter receptors. Our discoveries derived from two different experimental approaches. My finding originated from biochemical observation that cholesterol, used originally as a modifier of membrane fluidity, altered ligand binding to brain GABA$_A$ receptors. It culminated in the report describing modulation of brain GABA receptors by pregnenolone sulfate and some corticosteroids (Majewska *et al.*, 1985). On the other hand, Harrison and Simmons found steroid modulation on GABA receptor complex while studying the electrophysiological effects of the steroidal anesthetic, alphaxalone (Harrison and Simmons, 1984). Subsequently, while working together with Harrison at the National Institutes of Health, we made further discoveries employing combined neurochemical and electrophysiological techniques. We described potent barbiturate-like properties of reduced metabolites of progesterone and deoxycorticosterone (Majewska *et al.*, 1986) and determined structure–function relationships of agonistic steroid interactions with GABA$_A$ receptors (Harrison *et al.*, 1987).

In my subsequent research I, determined that steroids are bimodal modulators of GABA$_A$ receptors, some interacting with this receptor as agonists, others as antagonists (reviewed in Majewska, 1992). Among the most potent antagonists were pregnenolone sulfate (PS) (Majewska and Schwartz, 1987; Majewska *et al.*, 1988, 1990b) and dehydroepiandrosterone (DHEA) and its sulfate (DHEAS) (Majewska *et al.*, 1990a; Demirgoren *et al.*, 1991). Because in the mammalian brain GABA is the principal inhibitory neurotransmitter, the bi-directional regulation of its receptors by endogenous steroids is of paramount physiological significance, as concentrations of these steroids change during development and physiological states, such as stress, sleep, puberty, menopause, or ageing. Modulation of CNS neurotransmission by the steroids, originating either in the CNS or in periphery, links brain functions with animal's hormonal milieu, providing harmonious brain–body responses to internal and external stimuli.

DHEA and DHEAS (collectively termed DHEA/S) have been particularly exciting, though enigmatic, objects of interest for neuroscientists and endocrinologists. These two interconvertible steroids are not only the most abundant adrenal hormones in humans and other primates, but are also brain-borne neurosteroids (Baulieu et al., 1987), which, either in their native forms or as precursors of other steroids exert remarkable diversity of biological actions. They are widely regarded as youth-hormones, because their levels peak at early adulthood and then steadily decline with age (Orentreich et al., 1984). Some studies suggested correlations between blood levels of DHEA/S and vigor, resistance to cancers and cardiovascular diseases, and longevity (Barrett-Connor et al., 1986; Barrett-Connor and Edelstein, 1994) making this steroid a favorite object of interest for gerontobiologists. Our finding that DHEAS is a potent neuromodulator fueled interest in its multiple CNS actions.

This chapter describes the neuronal effects of DHEA/S and draws implications for its role in cognition, mental health, and ageing. Earlier reviews on biosynthesis, neurochemical, electrophysiological and behavioral actions of DHEAS have been published elsewhere (Majewska, 1992, 1995, 1999; Robel et al., 1999). Aspects of DHEA biosynthesis and metabolism are discussed in other chapters of this book (Mellon, Brown et al., Trincal et al.).

DHEAS AS EXCITATORY NEUROMODULATOR

The $GABA_A$ receptor was the first neurotransmitter receptor identified as a target for neurosteroid actions. It is a protein pentamer (Schofield et al., 1987), whose activation by GABA opens the associated chloride channel, leading to increased chloride conductance and usually to hyperpolarization of neuronal membrane. Heterogeneous forms of $GABA_A$ receptors exist in the brain, with different combinations of polypeptide subunits that change during development (Laurie et al., 1992; Poulter et al., 1992). This receptor is positively modulated by hypnotics and anesthetics, such as benzodiazepines and barbiturates, and negatively by convulsants such as picrotoxin or pentylenetetrazol.

We discovered that some neurosteroids regulate activity of the $GABA_A$ receptors as agonists and others as antagonists (reviewed in Majewska, 1992). DHEAS at low micromolar concentrations antagonizes function of this receptor and inhibits GABA-induced currents in neurons (Majewska et al., 1990; Demirgoren et al., 1991). This activity resembles the effects of pregnenolone (Majewska and Schwartz, 1987; Majewska et al., 1988), but the sites of action for these two steroids at the receptor appear to be distinct (Majewska et al., 1990b, 1991). The neurosteroids may be binding to hydrophobic pockets at the receptors, embedded in the plasma membrane (Majewska, 1992). The $GABA_A$-antagonistic features of DHEAS and PS contrast with $GABA_A$-agonistic profile of allopregnanolone (5α-pregnane-3α-ol, 20-one; also termed tetrahydro-progesterone, THP), tetrahydrodeoxycorticosterone (5α-pregnane-3α,21-diol-20-one, THDOC) (Majewska et al., 1986; Harrison et al., 1987), or androsterone (5α-androstane-3α-ol-17-one).

The GABA-active steroids modulate synaptic events and participate in brain development and neuronal plasticity. THP and THDOC are neuroinhibitory and they prolong inhibitory postsynaptic potentials (Harrison *et al.*, 1987), depress the depolarizing responses to glutamate, and block the glutamate-induced action potentials (Lambert *et al.*, 1990). In contrast, DHEAS and PS are neuroexcitatory, analeptic and proconvulsant (Heuser and Eidelberg, 1961; Carette and Poulain, 1984; Majewska *et al.*, 1989). The ubiquity of GABA$_A$ receptors in the CNS and their fundamental role in controlling neuronal excitability suggest that bimodal regulation of these receptors by the neurosteroids determines many brain functions and behaviors. Increase in synaptic concentrations of DHEAS or PS augments neuronal excitability and CNS arousal, whereas elevation of CNS levels of THP or THDOC enhances neuronal inhibition. Subsequently, steroids were found to also regulate other ionotropic neurotransmitter receptors. DHEA was reported as an agonistic modulator of the glutamatergic, NMDA receptor (Bergeron *et al.*, 1996), hence its global CNS effects are synergistically neuroexcitatory. A fine interplay may exist in the brain between the inhibitory (hypnotic/anxiolytic) steroids and the excitatory ones, which not only counteract each other's actions, but are also metabolically linked (Majewska, 1992), as excitatory steroids PS and DHEAS can be desulfated and converted to the inhibitory THP and androsterone.

DHEA AS HORMONE OF MENTAL HEALTH

DHEA and personality

Opposite neuromodulatory actions of different steroids suggest that their profile in plasma and in the CNS contributes to the manifestations of different neuropsychological states and personality traits. The steroid synthesis is genetically determined, but environmentally regulated and changes during development, ageing, pregnancy, stress, or disease. Theoretically, a higher proportion of excitatory steroids, such as DHEAS, to the inhibitory ones in the CNS may result in an anxiety-prone personality, typified by a higher level of resting arousal, greater sensitivity to incoming stimuli, a low threshold for positive hedonic tone (Eysenk, 1983) and a high sedation threshold (Lader, 1983). In contrast, a lower ratio of neuroexcitatory steroids to the inhibitory ones could result in a tendency for blunting reactions to stimuli, an extreme case of which would be a personality of "sensation seeker", typified by a very high threshold for arousal, requiring very strong stimuli to achieve positive hedonic tone. This concept was proposed in my earlier reviews (Majewska, 1987, 1992).

Indeed, empirical/clinical observations support this notion. Lower plasma levels of DHEAS have been found in individuals manifesting expansive or Type A personality than in those characterized as psychologically balanced type B (Hermida *et al.*, 1985; Fava *et al.*, 1987; Robel *et al.*, 1987). The type A personality was also marked by excessive secretion of cortisol, which may contribute to their greater cardiovascular morbidity and generally higher mortality rates. On the other hand, the individuals who had increased plasma levels of DHEAS resulting from congenital adrenal hyperplasia, were found to manifest anxiety disorder, which was successfully treated with ketoconazole, which reduces DHEA synthesis (Jacobs *et al.*, 1999).

DHEA and drug dependence

Sensation seeking personality is often linked to increased vulnerability for developing drug dependence. We have correlated plasma levels of DHEAS and cortisol with rate of relapse in recently treated cocaine addicts and found that refractory addicts had markedly lower levels of DHEAS than those who successfully recovered (Wilkins *et al.*, 1996; Majewska *et al.*, 2000). During cocaine discontinuation (3-week inpatient), which is typically accompanied by the activation of the HPA axis, the addicts who later proved to be successfully abstinent for at least 6 months manifested robust secretion of DHEAS, whereas the refractory addicts had low plasma levels of DHEAS and a generally blunted stress response. It thus appears that high blood levels of endogenous DHEAS may constitute a resilience factor, protecting cocaine addicts against relapse.

The biological significance of this phenomenon is unknown, but could be explained theoretically (Majewska and Wilkins, 2000). Increased synaptic activity of dopamine is believed to be at the core of the euphorigenic effects of psychostimulants such as cocaine and amphetamines (Pettit *et al.*, 1982). DHEAS as a neurostimulant potentiates secretion of dopamine (Murray and Gillies, 1997), thus it is conceivable that its high levels in the CNS functions as a hedonic substitute for cocaine and prevents relapse. We have tested the idea that DHEA might be an effective therapeutic agent for cocaine dependence and treated a group of addicts with this hormone (100 mg/day; 12-week double-blind trial). To our astonishment, DHEA not only was not therapeutic, but it even increased cocaine use and decreased patient's retention in the trial (Majewska *et al.*, 2000). From these studies, it is clear that DHEA/S plays a complex role in neurobiological substrates of hedonic and euphoric sensations and in drug addictions. It might be that higher blood levels of DHEA/S indeed are markers of resilience against drug dependence and other psychiatric disorders, but DHEA/S substitution in a vulnerable population is not therapeutic. In our case, DHEA treatment could have increased cocaine-induced euphoria, contributing to treatment failure. More research needs to be conducted to clarify this issue and to find out if deficient DHEAS *per se* plays any role in vulnerability to drug dependence, or if it is just an epiphenomenon of dysfunctional control of the hypothalamo-pituitary-axis, which contributes to susceptibility to drug addictions.

DHEA and aggression

Aggressivity is a complex psychobehavioral trait, determined genetically, neuroanatomically and environmentally. Generally, an individual's aggressivity remains under strong hormonal control. Androgens are typically proagressive in all species, whereas castration or treatment with anti-androgens results in gentling of males (Carlson, 1991). DHEAS has an ambiguous role in aggressiveness due to its multiple biological functions, as a weak androgen, an excitatory neuromodulator, and a precursor of other neuroactive steroids, such as strong androgens, estrogens, or anxiolytic androsterone. In animal experiments GABA-antagonists are usually proagressive (Puglisi-Allegra *et al.*, 1981), while GABA-agonists (Molina *et al.*, 1986; Kavaliers, 1988) have antiagressive effects. Thus DHEA/S as a GABA antagonist and androgen precursor would be expected to be proagressive. But DHEA was shown to decrease the aggressivity of castrated males against lactating females (Schlegel *et al.*, 1985). Because castration changes body metabolism of steroids, the anti-agressive effect of DHEA observed in castrates could possibly be mediated by its anxiolytic metabolite, androsterone, while its conversion to more potent proagressive androgens would be prevented.

We have evaluated the profile of DHEA/S secretion along with cortisol in cocaine-dependent humans, divided into those with or without history of violent aggression. We found that during cocaine abstinence the aggressive addicts had higher plasma levels of DHEAS and lower levels of cortisol than the non-aggressive (Buydens-Branchey *et al.*, 2001). These data are consistent with proagressive effects of DHEA/S. The specific role of DHEA in aggression might be complex however, because the individuals with Type A personality, whose aggressivity is oriented toward achievement rather than violence, appear to have rather low plasma concentrations of DHEA/S and high levels of cortisol. It is clear therefore that the contributions of adrenal steroids toward aggressive behaviors is complex and depends on many intertwined genetic, hormonal, neuroanatomical and environmental factors.

DHEA and affective disorders

DHEA/S as a direct $GABA_A$ antagonist and NMDA receptor agonist acts as a synergistic neurostimulant and analeptic. Indirectly, it also potentiates neuronal secretion of biogenic amines, including dopamine, noradrenaline and serotonin (Monnet *et al.*, 1995; Gillen *et al.*, 1999; Murray and Gillies, 1997). These effects resemble the action of antidepressant drugs and suggest that DHEA/S might have antidepressant properties. Some studies indeed showed that higher blood levels of DHEA/S were associated with a feeling of well being (Berr *et al.*, 1996; Cawood and Bancroft, 1996), whereas lower levels were reported in depressed and dysthymic patients (Ferrari *et al.*, 1997; Michael *et al.*, 2000), and patients with chronic fatigue syndrome (Kuratsune *et al.*, 1998). But other studies have not found a correlation between depression and low peripheral DHEA/S (Osran *et al.*, 1993), or even reported its higher levels in depressed individuals (Heuser *et al.*, 1998) and in refractory depressed patients, non responsive to ECT (Maayan *et al.*, 2000).

Such discrepancy may be due to heterogeneity and comorbidities of depressive disorders, and due to individual variations in adrenal responses to ACTH stimulation. For example, psychotic depression may be accompanied by increased DHEA/S production (Howard, 1992; Maayan *et al.*, 2000), similarly to schizophrenia (Oades and Schepker, 1994). Also, so-called "typical" depression is usually accompanied by overactive HPA axis, while "atypical depression" is characterized by a hypoactive HPA axis (Gold *et al.*, 1995). While most investigators use cortisol as an indicator of HPA activity, the stimulation of the HPA axis involves the secretion of many steroids with opposite psychoneurological effects and different kinetics of release (Genazzani *et al.*, 1998). DHEA/S is secreted in response to ACTH, but it is sensitive to lower plasma concentrations than cortisol (Arvat *et al.*, 2000), indicating that ACTH tonically stimulates DHEA/S release and that during mild stress this arousing, analeptic steroid is released by adrenals first, before cortisol and anxiolytic steroids such as pregnanolone and tetrahydrodeoxycorticosterone. DHEA/S is also secreted by a different zone of the adrenal cortex than cortisol (DHEA is produced by the zona reticularis and cortisol by zona fasciculata; Hornsby, 1995) and the relative proportions of both steroids depend on the activities of these two zones, which change during development and ageing. DHEA/cortisol ratios are high in young adults, but low in children and elderly. The situation is further complicated by the neuropsychological contribution of anxiolytic steroids released during stress, which counteract the effects of DHEA. Hence, it is extremely difficult to speculate about the psycho-biological significance of HPA axis activation in psychiatric disorders based on measures of a single steroid, while disregarding other components of the adrenal "hormonal soup", because in some individuals hyperactivity of the HPA axis may manifest itself by higher secretion of cortisol, while in others, by greater release of anxiogenic DHEA/S, or anxiolytic steroids.

Nonetheless, some studies showed antidepressant effects of administered DHEA in patients (Wolkowitz et al., 1997), and others reported its mood elevating (Morales et al., 1994) or mania-inducing (Dean, 2000) effects. Even more remarkable, emotional self-management training, which reduced negative psychological symptoms of stress, such as anxiety, burnout, hostility and guilt, while increasing positive measures such as vigor and caring, were found to be accompanied by a 100% increase of salivary DHEA/S and decrease of cortisol levels (Goodyer et al., 1996; McCraty et al., 1998). These data suggest that secretion of adrenal androgens and glucocorticoids, like many other physiological functions, can be controlled by the mind.

DHEA(S) IN MEMORY AND AGEING

DHEAS as an endogenous GABA antagonist and NMDA agonist was anticipated to improve learning and memory, because both neurotransmitter systems are critical for this function (reviewed in Majewska, 1992). DHEA/S may additionally potentiate memory by potentiating noradrenergic (Stanton and Survey, 1985; Monet et al., 1995), and cholinergic (Rhodes et al., 1996) activity in the brain. As predicted, DHEA/S augmented learning-linked hippocampal plasticity (Diamond et al., 1996), enhanced learning and memory when administered to rodents (Roberts et al., 1987; Flood et al., 1988; Shi et al., 2000), and potentiated attention and automatized memory in humans (Wolkowitz et al., 1995). Memory enhancing properties of DHEA may also be mediated by its effect of increasing rapid eye movement sleep (REM) (Friess et al., 1995), and by promoting neuronal sprouting (Roberts et al., 1987). All these effects suggest that DHEA may shape cognitive functions during development, puberty, pregnancy, stress, and ageing, when its blood levels change.

The body of evidence suggests that excessive GABAergic tone accelerates neuronal degeneration and brain ageing, as evidenced by cognitive impairments and cerebral atrophy observed in chronic users of benzodiazepines (Moodley et al., 1993). On the other hand, antagonism of GABA/benzodiazepine activity may slow down ageing (Marczynski et al., 1994). Ageing appears to be associated with amplification of GABA tone, as inferred from increased activity of glutamic acid decarboxylase in brains of aged animals (Marczynski et al., 1994) and from potentiated

GABA-activated currents in neurons from aged rats (Griffith and Murchison, 1995). Globally, such effects would reduce neuronal excitability, promote amnesia and neurodegeneration, while DHEA/S, as a GABA antagonist, may counteract these effects by improving cognition and facilitating neuronal survival (Roberts *et al.*, 1987; Flood and Roberts, 1988). DHEA was also shown to reduce neuronal injury resulting from ischemia (Aragno *et al.*, 2000; Li *et al.*, 2001), and to protect hippocampal neurons from excitatory aminoacid-induced toxicity at concentrations lower than those required for NMDA agonism (Kimonides *et al.*, 1998). These data suggest that DHEA/S might protect the brain from consequences of strokes, ischemic attacks, or other ageing-associated insults.

The results of clinical studies linking DHEA with degenerative diseases of ageing are ambiguous however. Some studies correlated lower plasma levels of DHEAS with Alzheimer's disease, multi-infarct dementia (Sunderland *et al.*, 1989; Nassman *et al.*, 1991), organic brain syndrome (Rudman *et al.*, 1990), and hippocampal perfusion deficits (Murialdo *et al.*, 2000). Others did not find such correlations (Carlson and Sherwin, 1999; Moffat *et al.*, 2000) and certain studies even reported that lower plasma levels of DHEA were associated with better cognitive functions in aged populations (Miller *et al.*, 1998; Morrison *et al.*, 2000).

DHEA/S AND SEXUAL DIMORPHISM

An intriguing sexual dimorphism was reported for DHEA/S concentrations and its postulated protective effects against diseases of ageing. Generally, women have lower plasma levels of DHEAS than men and different dynamics of DHEAS decline during ageing. While for both sexes the rate of decline of peripheral DHEAS is roughly parallel up to the age of 50, in later years the decline appears faster in men than in women (Orentreich *et al.*, 1984; Laughlin and Barrett-Connor, 2000). Some epidemiological studies found correlation of higher plasma levels of DHEA/S with lower incidence of cardiovascular diseases in men (Barrett-Connor and Goodman-Gruen, 1995; Feldman *et al.*, 1998), but in women there was either no association, or higher DHEA/S levels correlated with increased risk or cardiovascular diseases (Herrington *et al.*, 1990; Johannes *et al.*, 1999). Other studies reported lack of independent association between DHEAS and atherosclerosis in either sex (Kiechl *et al.*, 2000).

Postmortem studies conducted on a small number of individuals who died at age > 60 years, showed higher levels of DHEA/S in cerebral cortices of women than men (Lanthier and Patwardhan, 1986). This study needs to be replicated, but the difference might be significant, despite a small sample size, because it was consistent in four cerebral cortical areas, whereas there were no sex differences in cerebral concentrations of other steroids, and DHEA/S levels in other brain regions were not different between the two sexes. In the CSF of patients 50–60 years old, no sex difference was found in DHEA levels, but women had lower levels of DHEAS (515 pg/ml) than men (872 pg/ml) (Azuma *et al.*, 1993), and there was a negative correlation between CSF levels of DHEAS and ageing in men but not in women. DHEA/S levels in the CSF generally correlate with the plasma levels (Guazzo *et al.*, 1996), but might vary in neurodegenerative diseases and diseases with impaired blood brain barrier. Since increased concentrations of DHEAS were found in CSF of patients with neuropathies (Azuma *et al.*, 1993), higher levels of this steroid in the CSF of men than of women may not simply reflect their higher plasma levels, but may be also suggestive of greater loss of DHEAS from men's brains, as a result of their earlier ageing-related brain degeneration (Cowell *et al.*, 1994). Greater concentrations of DHEA/S in the CSF of men or patients with neuropathies may be also indicative of excessive oxidative stress (Brown *et al.*, 2000). More studies are needed to clarify the biological significance of sex difference in CSF levels of DHEAS and relationships between DHEAS and neurodegeneration.

DHEA/S REPLACEMENT

The theoretical premises of advantageous immune, cardiovascular, CNS and metabolic effects of DHEA suggested that its administration to ageing individuals might postpone geropathologies. Short term DHEA treatment was documented to be safe (Morales *et al.*, 1994; Legrain *et al.*, 2000). The initial placebo controlled efficacy trial documented positive hormonal, physical and psychological effects of 3-month treatment of middle-aged women and men with 50 mg of DHEA (Morales *et al.*, 1994). In a subsequent larger double-blind study, administration of 25 and 50 mg of DHEA to healthy older individuals for one year resulted in its moderate conversion to testosterone and estradiol and produced several positive effects on skin structure, bone metabolism and libido in women (Baulieu *et al.*, 2000). In other studies, both acute and 3-month administration of 50 and 100 mg of DHEA to older men also resulted in its biotransformation to estrogens and testosterone, but positive effects on well being and improved sexual function were not found (Arlt *et al.*, 1999a; Flynn *et al.*, 1999). Huppert *et al.* (2000) reviewed several double-blind trials with DHEA administration and reported lack of positive effects on well being and cognition in most of them.

In older women, or women with adrenal insufficiency, DHEA replacement leads to its dose-dependent conversion to estrogens and androgens, improvement of well-being (Arlt *et al.*, 1999a,b; Gebre-Medhin *et al.*, 2000) and increase of mineral bone density (Labrie *et al.*, 1998; Villareal *et al.*, 2000), but also has a negative effect on blood lipids, by reducing HDL cholesterol (Casson *et al.*, 1998; Arlt *et al.*, 1999b). The latter effect, along with elevation of testosterone, might be responsible for increased risk of heart diseases, observed in women with higher plasma DHEAS levels (Barrett-Connor and Goodman-Gruen, 1995; Johannes *et al.*, 1999).

Among other hormonal effects, the administration of DHEA increases serum levels of IGF-1 (Yen *et al.*, 1995; Arlt *et al.*, 1999b). By some, this might be viewed as a positive influence, responsible for anabolic and immune-stimulating effects of DHEA (Yen *et al.*, 1995), but it may have its dark site. DHEA, as a stimulant of IGF-1 and a precursor of estrogens and androgens may increase the risk of hormone-inducible cancers (Dorgan *et al.*, 1997; Stoll, 1999). Some animal studies suggested that DHEA inhibits carcinogenesis (Ratko *et al.*, 1991), but life-long treatment of mice with oral DHEA neither prolonged life, nor improved immunity or prevented ageing-related diseases, while increased incidence of cancers (Miller and Chrisp, 1999).

CONCLUSION

In contrast to exuberance generated by early studies with DHEA/S, portraying it as a fountain of youth and panacea for diseases of ageing, the body of recently accumulated evidence presents a humbler picture of this steroid. DHEA/S is clearly an active hormone and neuromodulator, which exerts many physiological effects. But its principal function is still unknown. The decline of adrenal production of DHEA/S during ageing looks more like a classic gero-epiphenomenon, similar to the decline of reproductive hormones, growth hormone, melatonin, etc. DHEA/S, like sex hormones, is not essential for post-natal life, as young children are healthy, happy, bright and resilient without it. It is more likely that DHEA/S serves as a supporter of gonads. While DHEA replacement benefits patients with adrenal insufficiency, there is as yet no convincing evidence that doing so in the elderly will protect them from diseases of ageing. The results of DHEA replacement studies in men and women showed modest, if any, beneficial physiological, psychological and cognitive effects, while producing negative effects on blood lipids and hormones, which may increase the risk of cardiovascular disorders and cancers. And there are no data on safety of long-term DHEA replacement therapy, which may eventually prove to do more harm than good, similar to hormone-replacement therapy in postmenopausal women (Schairer *et al.*, 2000). Such outcomes might be expected, because DHEA, like estrogens, testosterone, and IGF-1 are growth promoters and thus procancerogenic, particularly in older individuals. Animal studies suggest that reduction of IGF-1 activity prolongs life in several species (Hsin and Kenyon, 1999; Coschigano *et al.*, 2000), thus the IGF-1 enhancing effect of DHEA might negatively influence human longevity as well. Theoretically, DHEA may also shorten life by its thermogenic effects (Lardy *et al.*, 1995), which reduce lifespan in many species. Hence, in the absence of results from longitudinal DHEA replacement studies, the final verdict about benefits or harms of such therapies in humans is still pending.

REFERENCES

Aragno, M., Parola, S., Brignardello, E., Mauro, A., Tamagno, E., Manti, R. *et al.* (2000) Dehydroepiandrosterone prevents oxidative injury induced by transient ischemia/reperfusion in the rat brain of diabetic rats. *Diabetes*, **49**, 1924–1931.

Arlt, W., Callies, F., van Vlijmen, J.C., Koehler, I., Reincke, M., Bidlingmaier, M. *et al.* (1999b) Dehydroepiandrosterone replacement in women with adrenal insufficiency. *N. Engl. J. Med.*, **341**, 1013–1020.

Arlt, W., Haas, J., Callies, F., Reincke, M., Hubler, D., Oettel, M. *et al.* (1999a) Biotransformation of oral dehydroepiandrosterone in elderly men: significant increase in circulating estrogens. *J. Clin. Endocr. Metab.*, **84**, 2170–2176.

Arvat, E., Di Vito, L., Lanfranco, F., Maccario, M., Baffoni, C., Rosetto, R. *et al.* (2000) Stimulatory effect of adrenocorticotropin on cortisol, aldosterone, and dehydroepiandrosterone secretion in normal humans: dose-response study. *J. Clin. Endocr. Metab.*, **85**, 3141–3146.

Azuma, T., Matsubara, T., Shima, Y., Haeno, S., Fujimoto, T., Tone, K. *et al.* (1993) Neurosteroids in cerebrospinal fluid in neurologic disorders. *J. Neurol. Sci.*, **120**, 87–92.

Barrett-Connor, E. and Edelstein, S. (1994) A prospective study of dehydroepiandrosterone sulfate and cognitive function in an older population: the Rancho Bernardo study. *J. Am. Geriatr. Soc.*, **42**, 520–523.

Barrett-Connor, E. and Goodman-Gruen, D. (1995) The epidemiology of DHEAS and cardiovascular disease. *Ann. N. Y. Acad. Sci.*, **774**, 259–270.

Barrett-Connor, E., Khaw, K.T. and Yen, S.S.C. (1986) A prospective study of dehydroepiandrosterone sulfate, mortality, and cardiovascular disease. *N. Engl. J. Med.*, **315**, 1519–1524.

Baulieu, E.E., Robel, P., Vatier, O., Haug, A., Le Goascogne, C. and Bourreau, E. (1987) Neurosteroids: pregnenolone and dehydroepiandrosterone in the rat brain. In K. Fuxe and L.F. Agnati (eds), *Receptor–receptor interaction, a new intramembrane integrative mechanism*, Basingstoke: MacMillan, pp. 89–104.

Baulieu, E.E., Thomas, G., Legrain, S., Lahlou, N., Roger, M., Debuire, B. *et al.* (2000) Dehydroepiandrosterone (DHEA), DHEA sulfate, and aging: contribution of the DHEAge study to sociobiomedical issue. *Proc. Natl. Acad. Sci. USA*, **97**, 4279–4284.

Bergeron, R., De Montigny, C. and Debonnel, G. (1996) Potentiation of neuronal NMDA response induced by dehydroepiandrosterone and its suppression by progesterone: effects mediated via sigma receptors. *J. Neurosci.*, **16**, 1193–1202.

Berr, C., Lafont, S., Debuire, B., Lartigues, J.F. and Baulieu, E.E. (1996) Relationship of dehydroepiandrosterone sulfate in the elderly with functional, psychological, and mental status and short-term mortality: a French community-based study. *Proc. Natl. Acad. Sci. USA*, **93**, 13410–13415.

Brown, R.C., Cascio, C. and Papadopoulos, V. (2000) Pathways of neurosteroid biosynthesis in cell lines from human brain: regulation of dehydroepiandrosterone formation by oxidative stress and beta-amyloid peptide. *J. Neurochem.*, **74**, 847–859.

Buydens-Branchey, L., Branchey, M., Hudson, J. and Majewska, M.D. (2001) Perturbations of plasma cortisol and DHEA following discontinuation of cocaine use in cocaine addicts. *Psychoneuroendocrinology* (in press).

Carette, B. and Poulain, P. (1984) Excitatory effect of dehydroepiandrosterone, its sulfate ester and pregnenolone sulfate, applied by iontophoresis and pressure, on single neurons in the septo-optic area of the guinea pig. *Neurosci. Lett.*, **45**, 205–210.

Carlson, N.R. (1991) Aggressive Behavior. In *"Physiology of behavior"*, Allyn and Bacon, Boston. pp. 357–363.

Carlson, L.E. and Sherwin, B.B. (1999) Relationships among cortisol (CRT), dehydroepiandrosterone-sulfate (DHEAS), and memory in a longitudinal study of healthy elderly men and women. *Neurobiol. Aging*, **20**, 315–324.

Casson, P.R., Santoro, N., Elkind-Hirsch, K., Carson, S.A., Hornsby, P.J., Abraham, G. *et al.* (1998) Postmenopausal dehydroepiandrosterone administration increases free insulin-like growth factor-I and decreases high-density lipoprotein: a six month trial. *Fertil. Steril.*, **70**, 107–110.

Cawood, E.H. and Bancroft, J. (1996) Steroid hormones, the menopause, sexuality, and well-being of women. *Psycho. Med.*, **26**, 925–936.

Corpechot, C., Robel, P., Axelson, M., Sjovall, J. and Baulieu, E.E. (1981) Characterization and measurement of dehydroepiandrosterone sulfate in rat brain. *Proc. Natl. Acad. Sci. USA*, **78**, 4704–4707.

Corpechot, C., Synguelakis, M., Talha, S., Axelson, M., Sjovall, J., Vihko, R. *et al.* (1983) Pregnenolone and its sulfate ester in the rat brain. *Brain Research*, **270**, 119–125.

Coschigano, K.T., Clemmons, D., Bellush, L.L. and Kopchick, J.J. (2000) Assessment of growth parameters and life-span of GHR/BP gene disrupted mice. *Endocrinology*, **141**, 2608–2613.

Cowell, P.E., Turetsky, B.I., Gur, R.C., Grossman, R.I., Shtasi, D.L. and Gur, R.E. (1994) Sex difference in aging of the human frontal and temporal lobes. *J. Neurosci.*, **14**, 4748–4755.

Dean, C.E. (2000) Prasterone (DHEA) and mania. *Ann. Pharmacother.*, **34**, 1419–1422.

Demirgoren, S., Majewska, M.D., Spivak, C.E. and London, E.D. (1991) Receptor binding and electrophysiological effects of dehydroepiandrosterone sulfate, an antagonist of the GABA$_A$ receptor. *Neuroscience*, **45**, 127–135.

Diamond, D.M., Branch, B.J. and Fleshner, M. (1996) The neurosteroid dehydroepiandrosterone sulfate (DHEAS) enhances hippocampal primed burts, but not long-term potentiation. *Neurosci Lett.*, **202**, 204–208.

Dorgan, J.F., Stanczyk, F.Z., Longcope, C., Stephenson, H.E., Chang, L., Miller, R. *et al.* (1997) Relationship of serum dehydroepiandrosterone (DHEA), DHEAS sulfate, and 5-androsten-3-beta, 17 beta-diol to risk of breast cancer in postmenopausal women. *Cancer Epidemiol. Biomarkers Rev.*, **6**, 177–181.

Eysenk, H.J. (1983) Psychophysiology and personality: extraversion, neuroticism and psychoticism. In *Physiological correlates of human behaviour*, London, Academic Press, pp. 13–29.

Fava, M., Littman, A. and Halperin, P. (1987) Neuroendocrine correlates of the type A behavior pattern: a review and hypothesis. *Int. J. Psychiatry Med.*, **17**, 289–308.

Feldman, H.A., Johannes, C.B., McKinlay, J.B. and Longcope, C. (1998) Low dehydroepiandrosterone sulfate and heart disease in middle-aged men: cross sectional results from the Massachusetts male aging study. *Ann. Epidemiol.*, **8**, 217–128.

Ferrari, E., Locatelli, M., Arcaini, A. *et al.* (1997) Chronobiological study of some neuroendocrine features of major depression in elderly people. Abstracts, 79th *Annual Meeting of the Endocrine Society*. Endocrine Society Press., Bethesda, MD.

Flood, J.F. and Roberts, E. (1988) Dehydroepiandrosterone sulfate improves memory in aging mice. *Brain Res.*, **448**, 178–181.

Flood, J.F., Smith, G.E. and Roberts, E. (1988) Dehydroepiandrosterone and its sulfate enhance memory retention in mice. *Brain Res.*, **447**, 269–278.

Flynn, M.A., Weaver-Osterholtz, D., Sharpe-Timms, K.L., Allen, S. and Krause, G. (1999) Dehydroepiandrosterone replacement in aging humans. *J. Clin. Endocr. Metab.*, **84**, 1527–1533.

Friess, E., Trachsel, L., Guldner, J., Schier, T., Steiger, A. and Holsboer, F. (1995) DHEA administration increases rapid eye movement sleep and EEG power in the sigma frequency range. *Am. Physiol. Soc.*, **268**, E107–E113.

Gebre-Medhin, G., Husebye, E.S., Mallmin, H., Helstrom, L., Berne, C., Karlsson, F.A. *et al.* (2000) Oral dehydroepiandrosterone (DHEA) replacement therapy in women with Addison's disease. *Clin. Endocrinol.*, **52**, 775–780.

Genazzani, A.R., Petraglia, F., Bernardi, F., Casarosa, E., Salvestroni, C., Tonetti, A. *et al.* (1998) Circulating levels of allopregnanolone in humans: gender, age and endocrine influences. *J. Clin. Endocr. Metab.*, **83**, 2099–2103.

Gillen, G., Poryter, J.R. and Svec, F. (1999) Synergistic anorectic effect of dehydroepiandrosterone and d-fenfluramine on the obese Zucker rat. *Physiol Behav.*, **67**, 173–179.

Gold, P.W., Licinio, J., Wong, M.L. and Chrousos, G.P. (1995) Corticotropin releasing hormones in the pathophysiology of melancholic and atypical depression and in the mechanism of action of antidepressant drugs. *Ann. N. Y. Acad. Sci.*, **771**, 716–729.

Goodyer, I.M., Herbert, J., Atlham, P.M.E., Pearson, J., Secher, S.M. and Shiers, H.M. (1996) Adrenal secretion during major depression in 8- to 16-year-olds, I. Altered diurnal rhythms in salivary cortisol and dehydroepiandrosterone (DHEA) at presentation. *Psycho. Med.*, **26**, 145–156.

Griffith, W.H. and Murchison, D.A. (1995) Enhancement of GABA-activated membrane currents in aged Fisher 344 basal forebrain neurons. *J. Neurosci.*, **15**, 2407–2416.

Guazzo, E.P., Kirkpatrick, P.J., Goodyer, I.M., Shiers, H.M. and Herbert, J. (1996) Cortisol, dehydroepiandrosterone(DHEA), and DHEA sulfate in the cerebral fluid of man: relation to blood levels and the effects of age. *J. Clin. Endocr. Metab.*, **81**, 3951–3960.

Harrison, N.L. and Simmons, M.A. (1984) Modulation of GABA$_A$ receptor complex by steroid anesthetic. *Brain Res.*, **323**, 284–293.

Harrison, N.L., Majewska, M.D., Harrington, J.W. and Barker, J.L. (1987) Structure–activity relationships for steroid interaction with the τ-aminobutyric acid$_A$ receptor complex. *J. Pharm. Exp. Ther.*, **241**, 346–353.

Hermida, R.C., Halberg, F. and Del Pozo, F. (1985) Chronobiologic pattern discrimination of plasma hormone, notably DHEAS and TSH, classifies an expansive personality. *Chronobiologia*, **12**, 105–136.

Herrington, D.M., Gordon, G.B., Achuff, S.C., Trejo, J.F., Weisman, H.F., Kwiterovich, P.O. *et al.* (1990) Plasma dehydroepiandrosterone and dehydroepiandrosterone sulfate in patients undergoing diagnostic coronary angiography. *J. Am. Coll. Cardiol.*, **16**, 862–870.

Heuser, I., Deuschle, M., Lupp, P., Schweiger, U., Standhardt, H. and Weber, B. (1998) Increased diurnal plasma concentrations of dehydroepiandrosterone in depressed patients. *J. Clin. Endocr. Metab.*, **83**, 3130–3133.

Heuser, G. and Eidelberg, E. (1961) Steroid induced convulsions in experimental animals. *Endocrinology*, **69**, 915–924.

Hornsby, P.J. (1995) Biosynthesis of DHEAS by the human adrenal cortex and its age-related decline. *Ann. N. Y. Acad. Sci.*, **774**, 29–46.

Howard, J.S. (1992) Severe psychosis and the adrenal androgens. *Integr. Physiol. Behav. Sci.*, **27**, 209–215.

Hsin, H. and Kenyon, C. (1999) Signals from the reproductive system regulate the life-span of *C. elegans*. *Nature*, **399**, 362–366.

Huppert, F.A., Van Niekerk, J.K. and Herbert, J. (2000) Dehydroepiandrosterone replacement for cognition and well-being. *Cochrane Database Syst.*, Rev. CD000304.

Jacobs, A.R., Edelheit, P.B., Coleman, A.E. and Herzog, A.G. (1999) Late-onset congenital adrenal hyperplasia: a treatable cause of anxiety. *Biol. Psychiatry*, **46**, 856–859.

Johannes, C.B., Stellato, R.K., Feldman, H.A., Longcope, C. and McKinlay, J.B. (1999) Relation of dehydroepiandrosterone and dehydroepiandrosterone sulfate with cardiovascular disease risk factors in women: longitudinal results from Massachsetts women's halth study. *J. Clin. Epidemiol.*, **52**, 95–103.

Kavaliers, M. (1988) Inhibitory influences of adrenal steroid, 3α,5α-tetrahydrodeoxycorticosterone on aggression and defeat-induced analgesia in mice. *Psychopharmacol.*, (Berlin). **95**, 488–492.

Kiechl, S., WilleIt, J., Bonora, E., Schwarz, S. and Xu, Q. (2000) No association between dehydroepiandrosterone sulfate and development of atherosclerosis in a prospective population study (Bruneck Study). *Artherioscler. Thromb. Vasc. Biol.*, **20**, 1094–1100.

Kimonides, V.G., Khatibi, N.H., Svendsen, C.N., Sofroniew, M.V. and Herbert, J. (1998) Dehydroepiandrosterone (DHEA) and DHEA-sulfate (DHEAS) protect hippocampal neurons against excitatory amino acid-induced neurotoxicity. *Proc. Natl. Acad. Sci. USA*, **95**, 1852–1857.

Kuratsune, H., Yamaguti, K., Sawada, M., Kodate, S., Machii, T., Kanakura, Y. *et al.* (1998) Dehydroepiandrosterone sulfate deficiency in chronic fatigue syndrome. *Int. J. Mol. Med.*, **1**, 143–146.

Labrie, F., Belanger, A., Luu-The, V., Labrie, C., Simard, J., Cusan, L. *et al.* (1998) DHEA and the intracrine formation of androgens and estrogens in peripheral target tissues: its role during aging. *Steroids*, **63**, 322–328.

Lader, M. (1983) Anxiety and depression. In A. Gale and J.A. Edwards (eds). *Physiological correlates of human behaviour*, London: Academic Press, pp. 155–167.

Lambert, J.J., Peters, J.A., Strugges, N.C. and Hales, T.G. (1990) Steroid modulation of the GABA-A receptor complex: electrophysiological studies. In *Steroid and neuronal activity*, Ciba Foundation Symposium, John Wiley and Sons, Chichester, **153**, pp. 56–70.

Lanthier, A. and Patwardhan, V.V. (1986) Sex steroids and 5-en-3β-hydroxysteroids in specific regions of the human brain and cranial nerves. *J. Steroid Biochem.*, **25**, 445–449.

Lardy, H., Kneer, N., Bellei, M. and Bobyleva, V. (1995) Induction of thermogenic enzymes by DHEA and its metabolites. *Ann. N. Y. Acad. Sci.*, **774**, 171–179.

Laughlin, G.A. and Barrett-Connor, E. (2000) Sexual dimorphism in the influence of advanced aging on adrenal hormone levels: the Rancho Bernardo study. *J. Clin. Endocr. Metab.*, **85**, 3561–3568.

Laurie, D.J., Wisden, W. and Seeburg, P. (1992) The distribution of thirteen GABA$_A$ receptor subunit mRNAs in the rat brain. III. Embryonic and postnatal development. *J. Neurosci.*, **12**, 4151–4172.

Legrain, S., Massien, C., Lahlou, N., Roger, M., Debuire, B., Diquet, B. *et al.* (2000) Dehydroepiandrosterone replacement administration: pharmaco-kinetic and pharmacodynamic studies in healthy elderly subjects. *J. Clin. Endocr. Metab.*, **85**, 3208–3217.

Li, H., Klein, G., Sun, P. and Buchan, A.M. (2001) Dehydroepiandrosterone (DHEA) reduces neuronal injury in rat model of global cerebral ischemia. *Brain Res.*, **888**, 263–266.

Maayan, R., Yagorowski, Y., Grupper, D., Weiss, M., Shraif, B., Kaoud, M.A. *et al.* (2000) Basal plasma dehydroepiandrosterone sulfate level: a possible predictor of response to electroconvulsive therapy in depressed psychotic patients. *Biol. Psychiatry*, **48**, 693–701.

Majewska, M.D. (1987) Steroids and brain activity. Essential dialogue between body and mind. *Biochem. Pharmacol.*, **36**, 3781–3788.

Majewska, M.D. (1992) Neurosteroids: Endogenous bimodal modulators of the GABA$_A$ receptor. Mechanism of action and physiological significance. *Progr. Neurobiol.*, **38**, 279–295.

Majewska, M.D. (1995) Neuronal Actions of Dehydroepiandrosterone. Possible Roles in Brain Development, Aging, Memory and Affect. In *Dehydroepiandrosterone (DHEA) and aging. Ann. N. Y. Acad. Sci.*, **774**, 111–120.

Majewska, M.D. (1999) Neurosteroid Antagonists of the GABA$_A$ Receptors. In E.E. Baulieu, P. Robel and M. Schumacher, (eds), *Neurosteroids. A new regulatory function in the nervous system*, Humana Press, Totowa, US, pp. 155–166.

Majewska, M.D., Bisserbe, J.C. and Eskay, R.E. (1985) Glucocorticoids are modulators of the GABA$_A$ receptors in brain. *Brain Res.*, **339**, 178–182.

Majewska, M.D., Bluet-Pajot, M.T., Robel, P. and Baulieu, E.E. (1989) Pregnenolone sulfate antagonizes barbiturate-induced sleep. *Pharmacol. Biochem. Behav.*, **33**, 701–703.

Majewska, M.D., Demirgoren, S. and London, E.D. (1990b) Binding of pregnenolone sulfate to rat brain membranes suggests multiple sites of steroid action at the GABA$_A$ receptor. *Eur. J. Pharmacol. (Molec. Pharmacol. Sect.)*, **189**, 307–315.

Majewska, M.D., Demirgoren, S., Spivak, C.E. and London, E.D. (1990a) The neurosteroid dehydroepiandrosterone sulfate is an antagonist of the GABA$_A$ receptor. *Brain Res.*, **526**, 143–146.

Majewska, M.D., Mienville, J.M. and Vicini, S. (1988) Neurosteroid pregnenolone sulfate antagonizes electrophysiological responses to GABA in neurons. *Neurosci. Lett.*, **90**, 279–284.

Majewska, M.D., Harrison, N.L., Schwartz, R.D., Barker, J.L. and Paul, S. (1986) Steroid hormone metabolites are barbiturate-like modulators of the GABA receptor. *Science*, **232**, 1004–1007.

Majewska, M.D. and Schwartz, R.D. (1987) Pregnenolone-sulfate: an endogenous antagonist of the τ-aminobutyric acid receptor complex in brain? *Brain Res.*, **404**, 355–360.

Majewska, M.D., Spivak, C.E. and London, E.D. (1991) Receptor binding and electrophysiological effects of dehydroepiandrosterone sulfate, an antagonist of the GABA$_A$ receptor. *Neuroscience*, **45**, 127–135.

Majewska, M.D. and Wilkins, J. (2000) Low plasma levels of DHEAS as markers for refractory cocaine dependence. *Neuropsychopharmacology*, **23**, S 46.

Majewska, M.D., Wilkins, J., Shoptaw, S., Ling, W., van Gorp, W.G. and Li, S.H. (2000) DHEAS: A Maverick Agent in Cocaine Dependence. *Am. Society for Neurosci.* November 2000.

Marczynski, T.J., Artwohl, J. and Marczynski, B. (1994) Chronic administration of flumazenil increases life span and protects rats from age-related loss of cognitive functions: a benzodiazepine/GABAergic hypothesis of brain aging. *Neurobiol. Aging*, **15**, 69–84.

McCraty, R., Barrios-Choplin, B., Rozman, D., Atkinson, M. and Watkins, A.D. (1998) The impact of a new emotional self-management program on stress, emotions, heart rate variability, DHEA and cortisol. *Integr. Physiol. Behav. Sci.*, **33**, 151–170.

Michael, A., Jenaway, A., Paykel, E.S. and Herbert, J. (2000) Altered salivary dehydroepiandrosterone levels in major depression in adults. *Biol. Psychiatry*, **48**, 989–995.

Miller, R.A. and Chrisp, C. (1999) Lifelong treatment with oral DHEA sulfate does not preserve immune function, prevent disease, or improve survival in genetically heterogeneous mice. *J. Am. Geriatr. Soc.*, **47**, 960–966.

Miller, T.P., Taylor, J., Rogerson, S., Mauricio, M., Kennedy, Q., Schatzberg, A. *et al.* (1998) Cognitive and noncognitive symptoms in dementia patients: relationship to cortisol and dehydroepiandrosterone. *Int. Psychogeriatr.*, **10**, 85–96.

Moffat, S.D., Zonderman, A.B., Harman, S.M., Blackman, M.R., Kawas, C. and Resnick, S.M. (2000) The relationship between longitudinal declines in dehydroepiandrosterone sulfate concentration and cognitive performance in older men. *Arch. Intern. Med.*, **160**, 2193–2198.

Molina, V., Ciesielski, L., Gobaille, S. and Mandel, P. (1986) Effects of potentiation of the GABAergic neurotransmission in the olfactory bulbs on mouse-killing behavior. *Pharmacol. Biochem. Behav.*, **24**, 657–664.

Monnet, F.P., Mahe, V., Robel, P. and Baulieu, E.E. (1995) Neurosteroids, via sigma receptors, modulate the [^{3}H]norepinephrine release evoked by *N*-methyl-D-aspartame in the rat hippocampus. *Proc. Natl. Acad. Sci. USA*, **92**, 3774–3778.

Moodley, P., Golombek, S., Shine, P. and Lader, M. (1993) Computerized axial brain tomograms in long-term benzodiazepine users. *Psychiatry Res.*, **48**, 135–144.

Morales, A.J., Nolan, J.J., Nelson, J.C. and Yen, S.S.C. (1994) Effect of replacement dose of dehydroepiandrosterone in men and women of advancing age. *J. Clin. Endocr. Metab.*, **78**, 1360–1367.

Morrison, M.F., Redei, E., TenHave, T., Parmelee, P., Boyce, A.A., Sinha, P.S. *et al.* (2000) Dehydroepiandrosterone sulfate and psychiatric measures in a frail, elderly residential care population. *Biol. Psychiatry*, **47**, 144–150.

Murialdo, G., Nobili, F., Rollero, A., Gianelli, M.V., Copello, F., Rodriguez, G. *et al.* (2000) Hippocampal perfusion and pituitary-adrenal axis in Alzheimer's disease. *Neuropsychobiology*, **42**, 51–57.

Murray, H.E. and Gillies, G.E. (1997) Differential effects of neuroactive steroids on somatostatin and dopamine secretion from primary hypothalamic cell cultures. *J. Neuroendocrinol.*, **9**, 387–395.

Nassman, B., Olsson, T., Backstrom, T., Eriksson, S., Grakvist, K., Viitanen, M. *et al.* (1991) Serum dehydroepiandrosterone sulfate in Alzheimer's disease and multi-infarct dementia. *Biol. Psychiatry*, **30**, 684–690.

Oades, R.D. and Schepker, R. (1994) Serum gonadal steroid hormones in young schizophrenic patients. *Psychoneuroendocrinology*, **19**, 373–385.

Orentreich, N., Brind, J.L., Ritzler, R.L. and Vogelman, J.H. (1984) Age changes and sex differences in serum dehydroepiandrosterone sulfate concentrations throughout adulthood. *J. Clin. Endocr. Metab.*, **59**, 551–555.

Osran, H., Reist, C., Chen, C.C., Lifrak, E.T., Chicz-DeMet, A. and Parker, L.N. (1993) Adrenal androgens and cortisol in major depression. *Am. J. Psychiatry*, **150**, 806–809.

Pettit, H.O., Ettenberg, A., Bloom, E.E. and Koob, G.F. (1982) Destruction of dopamine in the nucleus accumbens selectively attenuates cocaine but not heroin self-administration in rats. *Psychopharmacology*, **84**, 167–173.

Poulter, M.O., Barker, J.L., O'Carroll, A.M., Lolait, S.J. and Mahan, L.C. (1992) Differential and transient expression of GABA$_A$ receptor alpha-subunit MRNAS in the developing rat CNS. *J. Neurosci.*, **12**, 2888–2900.

Puglisi-Allegra, S., Simler, S., Kempf, E. and Mandel, P. (1981) Involvement of the GABA-ergic system on shock-induced aggressive behavior in two strains of mice. *Pharmacol. Biochem. Behav.*, **14**, Suppl. 1, 13–18.

Ratko, T.A., Detrisac, C.J., Mehta, R.G., Kellof, G.J. and Moon, R.C. (1991) Inhibition of rat mammary gland chemical carcinogenesis by dietary dehydroepiandrosterone or fluorinated analogue of dehydroepiandrosterone. *Cancer Res.*, **51**, 481–486.

Rhodes, M.E., Li, P.K., Flood, J.F. and Johnson, D.A. (1996) Enhancement of hippocampal acetylcholine release by the neurosteroid dehydroepiandrosterone sulfate: an *in vivo* microdialysis. *Brain Res.*, **733**, 284–286.

Robel, P., Baulieu, E.E., Synguelakis, M. and Halberg, F. (1987) Chronobiologic dynamics of delta5-3beta-hydroxysteroids and glucocorticoids in rat brain and plasma and human plasma. *Prog. Clin. Biol. Res.*, **227A**, 451–465.

Robel, P., Schumacher, M. and Baulieu, E.E. (1999) Neurosteroids: from definition and Biochemistry to Psychopathologic Function. In E.E. Baulieu, P. Robel and M. Schumacher (eds), *Neurosteroids. A new regulatory function in the nervous system*, Humana Press, Totowa, US, pp. 1–26.

Roberts, E., Bologa, L., Flood, J.F. and Smith, G.E. (1987) Effects of dehydroepiandrosterone and its sulfate on brain tissues in culture and on memory in mice. *Brain Res.*, **406**, 357–362.

Rudman, D., Shetty, K.R. and Mattson, D.E. (1990) Plasma dehydroepiandrosterone sulfate in nursing home men. *J. Am. Geriatr. Soc.*, **38**, 421–427.

Schairer, C., Lubin, J., Troisi, R., Sturgeon, S., Brinton, L. and Hoover, R. (2000) Menopausal estrogen and estrogen-progestin replacement therapy and breast cancer risk. *J.A.M.A.*, **283**, 485–491.

Schlegel, M.L., Spetz, J.F., Robel, P. and Haug, M. (1985) Studies on the effects of dehydroepiandrosterone and its metabolites on attack by castrated mine on lactating intruders. *Physiol. Behav.*, **34**, 867–870.

Schofield, P.R., Darlison, M.G., Fujita, M., Burt, D.R., Stephenson, F.A., Rodriguez, H. *et al.* (1987) Sequence and functional expression of the GABA$_A$ receptor shows a ligand-gated receptor super-family. *Nature*, **328**, 8221–8227.

Shi, J., Schulze, S. and Lardy, H.A. (2000) The effect of 7-oxo-DHEA acetate on memory in young and old C57/BL/6 mice. *Steroids*, **65**, 124–129.

Stanton, P.K. and Survey, J.M. (1985) Depletion of norepinephrine, but not serotonin, reduces long-term potentiation in dentate gyrus of rat hippocampal slices. *J. Neurosci.*, **5**, 2169–2176.

Stoll, B.A. (1999) Dietary supplements of dehydroepiandrosterone in relation to breast cancer risk. *Eur. J. Clin. Nutr.*, **53**, 771–775.

Sunderland, T., Merril, C.R., Harrington, M.G., Lawlor, B.A., Molchan, S.E., Martinez, R. *et al.* (1989) Reduced plasma dehydroepiandrosterone concentrations in Alzheimer's disease. *Lancet*, **8662**, 570.

Villareal, D.T., Holloszy, J.O. and Kohrt, W.M. (2000) Effect of DHEA replacement on bone mineral density and body composition in elderly women and men. *Clin. Endocrinol.*, **53**, 561–568.

Wilkins, J., Van Gorp, W., Hinken, C., Welch, B., Wheatley, S., Plotkin, D. *et al.* (1997) Relapse to Cocaine Use May be Predicted in Early Abstinence By Low Plasma Dehydroepiandrosterone. *NIDA Research Monograph*, **174**, 277.

Wolkowitz, O.M., Reus, V.I., Keebler, A., Nelson, N., Friedland, M., Brisendine, L. *et al.* (1997) Dehydroepiandrosterone (DHEA) treatment of depression. *Biol. Psychiatry*, **41**, 311–318.

Wolkowitz, O.M., Reus, V.I., Roberts, E., Manfredi, F., Chan, T., Ormitson, S. *et al.* (1995) Antidepressant and cognition-enhancing effects of DHEA in major depression. *Ann. N. Y. Acad. Sci.*, **774**, 337–339.

Yen, S.S., Morales, A.J. and Khorram, O. (1995) Replacement of DHEA in aging men and women. Potential remedial effects. *Ann. N. Y. Acad. Sci.*, **774**, 128–142.

Chapter 5

DHEA effects on liver

Doris Mayer

INTRODUCTION

Dehydroepiandrosterone (5-androsten-3β-ol-17-one; DHEA) and its sulfate ester dehydroepiandrosterone sulfate (DHEAS) are the principal steroids secreted by the adrenal cortex and circulating in the blood in humans (Baulieu, 1962; Schwartz *et al.*, 1988). For a long time DHEA(S) represented a hormone in search of a function; but there is now increasing evidence that DHEA(S), besides its function as a precursor in the biosynthesis of potent androgens and estrogens, has effects in various parts of the body, particularly on the immune system and the cardiovascular system (summarized in Kalimi and Regelson, 2000), and in the brain where DHEA acts as a neurosteroid (Baulieu and Robel, 1996). Although significant progress has been achieved during recent years (Bellino *et al.*, 1995; Kalimi and Regelson, 2000), the mechanism of DHEA action has not yet been fully elucidated.

DHEA has been reported to exert a number of beneficial effects in experimental animals, including decrease in body weight and fat deposits, anti-ageing, anti-diabetic and anti-glucocorticoid effects, and stimulation of the immune system (reviewed in Kalimi and Regelson, 1990, 2000; Bellino *et al.*, 1995). DHEA has also been attributed an anti-cancer effect since it can inhibit chemically induced carcinogenesis in some animal models (reviewed in Mayer, 1998; Feo *et al.*, 2000). Conversely, DHEA shows unfavorable effects in rat liver such as hepatomegaly and peroxisome proliferation, and after long-term treatment hepatocellular carcinomas occur (Rao *et al.*, 1992b,c; Metzger *et al.*, 1995; Mayer, 1998; Mayer *et al.*, 1998b, 2000; Bannasch *et al.*, 2000).

The main functions of the liver are gluconeogenesis and the maintenance of glucose homeostasis in the blood thus providing other organs including the brain with glucose, degradation and synthesis of fatty acids, elimination of nitrogen by production of urea, biosynthesis of cholesterol and bile acids, and metabolism and elimination of drugs and endogenous compounds such as steroid hormones. The present review will summarize mainly the effects of DHEA on metabolic processes observed in rat liver of both genders upon short and long-term administration of DHEA in the diet. Findings on other species including humans will be included where appropriate.

PHARMACOKINETICS, UPTAKE AND METABOLISM OF DHEA(S)

The site of DHEA production has not been clarified in the rat. According to van Werden *et al.* (1992), the adrenal glands do not produce DHEA in the rat. Nevertheless, DHEA and DHEAS are present in the blood at low levels under physiological conditions (Hamilton *et al.*, 1991; Hobe *et al.*, 1994a). In rats treated with 0.3% or 0.6% DHEA (w/w) in the diet, blood concentrations of DHEA(S) increased significantly, particularly in females (Hobe *et al.*, 1994a,b). While the plasma levels of DHEA(S) studied in males over the time period of 8 days were below 1 µg/ml, the levels of DHEA were higher by factors of 32 to 38 in females 48 h after dietary administration of DHEA at concentrations of 0.3% or 0.6%, respectively. Estradiol and testosterone levels in the blood were significantly increased in female rats after 3 days of treatment, in males only estrogen levels were elevated (K. Kopplow and D. Mayer, unpublished).

Uptake studies with isolated rat hepatocytes have shown that DHEA enters the cells rapidly by diffusion while DHEAS requires a transport system (Reuter and Mayer, 1995). The transport system in rats was characterized as a saturable ($K_m = 17 \mu M$) and energy dependent carrier of the multispecific bile acid transport type (Reuter and Mayer, 1995), and has been suggested to be strongly related to the organic anion transporting polypeptide, a multispecific steroid carrier, in humans (Kullak-Ublick *et al.*, 1998).

In comparison to the numerous reports on the effects of DHEA in rats, only a few studies on its metabolism have been published. Steroid metabolism and excretion show significant differences in males and females. In the male rat, hydroxylation reactions dominate, whereas the female rat is characterized by a more hydrogenating steroid metabolism (Starka and Kutova, 1962; Schriefers *et al.*, 1971). One hour after oral administration of a single dose of radioactively labeled DHEA (10 mg/kg body weight) to rats, most of the radioactivity was found in the liver and kidney (Hobe *et al.*, 1994a,b). After 24 h, radioactivity was mainly retained in the pituitary gland and bone marrow (Hobe *et al.*, 1994b). Male rats metabolized and eliminated DHEA at a significantly higher rate than females. Radioactivity was mainly eliminated in the faeces in male rats and in the urine in females. Metabolite patterns in plasma, bile and urine following administration of [³H]-DHEA (10 mg/kg body weight) were assayed by thin layer chromatography and densitometry. Mainly polar compounds were excreted in male rats, whereas females converted DHEA to less polar metabolites, some of them having estrogenic or androgenic activity (Hobe *et al.*, 1994a,b, 2000). The pattern of metabolites produced from [³H]-DHEA by rat liver homogenates *in vitro* was investigated by the same methods yielding similar results. 5-Androstene-3β,17β-diol (ADIOL) was the principal metabolite in both sexes followed by 7-hydroxy DHEA and 7-oxo-DHEA in males; the latter compounds were found only at low levels in females.

No sex differences in DHEA-biotransformation could be detected with liver homogenates and in urine from male and female NMRI mice. ADIOL represented the predominant metabolite also in mice. 7α- and 7β-hydroxy DHEA and 5-androstene-3β,7α,17β-triol and other compounds even more polar than these were observed in urine from mice. This agrees with findings of Rose *et al.* (1997)

and Doostzadeh *et al.* (1998) who described that hydroxylation in the 7-position of DHEA plays a predominant role in mice. Androsterone and etiocholanolone which represent the main DHEA metabolites in humans do not seem to be produced in rodents (Bradlow and Zumoff, 2000; Hobe *et al.*, 2000)

ALTERATIONS IN ENERGY METABOLISM

DHEA has been described by many authors to reduce the body weight of normal and particularly of obese diabetic rats and mice without lowering food intake (Yen *et al.*, 1977; reviewed by Cleary, 1990, 2000; Bradlow and Zumoff, 2000; McIntosh, 2000; Svec and Porter, 2000). In addition to causing weight loss and reduction of body fat in obese animals, oral administration of DHEA results in a decrease of fasting levels of serum triglycerides, glucose and insulin (summarized in Bradlow and Zumoff, 2000; McIntosh, 2000). DHEA increased resting metabolic rate and whole body respiration, which could be explained by changes of serum thyroxine (T4) levels and T4 5′-deiodinase activities in liver and pituitary gland in DHEA-treated rats (McIntosh and Berdanier, 1991, 1992). Thyroidectomized rats did not respond to DHEA (Song *et al.*, 1989). It was concluded that DHEA alters the thyroid hormone status and affects energy balance by increasing oxygen consumption (McIntosh and Berdanier, 1992). Increased energy expenditure would result in less energy for fat synthesis and storage, thereby reducing body fat. Weight loss after DHEA administration is seen only with oral DHEA, no effect was seen when DHEA was given intraperitoneally (Lardy *et al.*, 1998). This suggests that metabolism of DHEA in either the gut or the liver is involved in the weight-loss effects (Bradlow and Zumoff, 2000).

Interest in the liver as a potential site of DHEA's antiobesity effect resulted from the observations that livers from DHEA-treated rats were darker in color with little fatty infiltration, and liver weights were elevated either absolutely or relative to body weight compared to livers from non-treated rats (Cleary *et al.*, 1983, 1984b; Mayer *et al.*, 1988; Cleary, 1990; Metzger *et al.*, 1995). The higher liver weights were accompanied by elevated DNA, RNA and protein levels following long-term DNA treatment (Cleary *et al.*, 1984b, 1985). Elevated mitochondrial protein was observed as early as three days after initiation of DHEA-administration (Mohan and Cleary, 1991). Other early effects of DHEA treatment in rat liver were peroxisome proliferation and induction of enzymes of peroxisomal β-oxidation (Frenkel *et al.*, 1990; Prough *et al.*, 1994), increase in lipid peroxidation (Swierczynski *et al.*, 1996), and alterations in activities of key enzymes of carbohydrate metabolism and lipogenesis (Mayer *et al.*, 1988, 1996, 1998a).

Carbohydrate metabolism

In recent studies, we have analysed the effect of long and short-term administration of DHEA (0.6% in the diet) to male and female Sprague-Dawley rats on glycogen content and on the activity of key enzymes of carbohydrate metabolism in the liver (Mayer *et al.*, 1988, 1996, 1998a). Changes in glucose metabolism were observed as early as three days after initiation of DHEA-treatment. Biochemical assays revealed a significant reduction of the glycogen content of the livers to 40–60%

of controls in both genders. Glycogen content remained low during the whole period (up to 20 weeks) of treatment. Female rats were slightly more sensitive than males. Activities of the glycogen metabolizing enzymes synthase and phosphorylase were also reduced. We also observed a persistent decrease, in both sexes, in the activity of key enzymes of glycolysis (glucokinase, hexokinase, pyruvate kinase) and gluconeogenesis (glucose-6-phosphatase, fructose-1,6-bisphosphatase) with the exception that gluconeogenic enzymes were significantly increased after 3 days of treatment. The glycolytic enzymes showed a stronger reduction in activity than the gluconeogenic enzymes. Glucokinase and phosphorylase mRNA levels were found to be significantly decreased in livers of male rats treated for 14 days. These observations indicate that DHEA equally affects opposite pathways of glucose metabolism, such as glycolysis and gluconeogenesis. The loss of enzyme activity seems to be due to downregulation of gene expression of key enzymes of glucose metabolism which suggests implication of DHEA in central regulatory mechanisms rather than direct inhibitory effects on enzyme activities. The observation that the rate-limiting key enzymes of metabolic pathways are affected by DHEA-treatment suggests that the fluxes through the respective pathways are altered. These observations agree with findings on hepatocytes isolated from DHEA-treated rats which showed a reduced rate of gluconeogenesis compared to hepatocytes isolated from control rats (McIntosh, 2000). It was concluded that a reduced gluconeogenesis may contribute to the decrease in blood glucose levels observed in diabetic rats after DHEA-treatment (reviewed by Cleary, 1990) and to a lower extent in normal female rats (Mayer *et al.*, 1996). In contrast, glucocorticoids increased blood glucose by stimulating hepatic gluconeogenesis. Therefore, DHEA was attributed antiglucocorticoid properties.

Glucose-6-phosphate dehydrogenase (G6PDH), the key enzyme of the pentose phosphate pathway, is the rate limiting enzyme in the *de novo* synthesis of ribose required for cell proliferation, and the main source for NADPH, a cofactor necessary for biosynthetic processes. As to the effect of DHEA-treatment on G6PDH activity, conflicting results were reported ranging from decrease (Cleary *et al.*, 1983, 1984a; Shepherd and Cleary, 1984; Tagliaferro *et al.*, 1986; Garcea *et al.*, 1987, 1988; Mayer *et al.*, 1988) over no change (Shepherd and Cleary, 1984; Casazza *et al.*, 1986) to increase (Shibata *et al.*, 1995; Mayer *et al.*, 1998). The finding whether DHEA increases or decreases G6PDH activity in rat liver seemed to be dependent on the experimental schedule, on the length of treatment and the rat strain chosen. Decreased G6PDH activity was usually observed in obese rats, and G6PDH elevation observed after fasting/refeeding was prevented by DHEA. In our own experiments, we observed an increase of G6PDH activity in male and no change in female Sprague Dawley-rats after short-term treatment (≤4 weeks) (Mayer *et al.*, 1996). After long-term treatment (≥20 weeks) G6PDH activity returned to control levels or was slightly reduced in males and was elevated in females (Mayer *et al.*, 1998a).

Lipid metabolism

G6PDH is not only the rate-limiting enzyme for ribose *de novo* biosynthesis, but also a key enzyme for lipogenesis. DHEA has been shown to be an uncompetitive inhibitor *in vitro* of G6PDH with a $K_i \approx 18\,\mu M$ (Marks and Banks, 1960). The antiobesity action of DHEA first described by Yen *et al.* (1977) and later reproduced by a number of researchers (see Cleary, 1990, 2000; Bradlow and Zumoff, 2000) was initially explained by inhibition of G6PDH activity. Theoretically, inhibition of G6PDH would result in a decreased production of NADPH required for lipid biosynthesis. However, it is doubtful if DHEA concentrations as high as 18 μM are ever attained in liver. Hobe *et al.* (1994a, 2000) reported DHEA plasma concentrations of ≤1 μM in rats treated with 0.6% DHEA in the diet. DHEAS which was found at much higher concentrations in female rats (Hobe *et al.*, 1994a) does not inhibit G6PDH. It was also reported that the level of NADPH (Casazza *et al.*, 1986; Garcea *et al.*, 1988) and the NADPH/NADP-ratio (Casazza *et al.*, 1986) were unchanged in DHEA-treated liver. Cleary and co-workers observed no differences in metabolism of glucose by the pentose phosphate pathway in isolated hepatocytes prepared from rats following either short-term or long-term treatment with DHEA in comparison to non-treated rats (Diersen-Schade and Cleary, 1989; Finan and Cleary, 1988). Also, the fact that DHEA-treatment resulted in a strong elevation of malic enzyme activity (Mayer *et al.*, 1996, 1998a) and a less pronounced increase in isocitrate dehydrogenase activity (Garcea *et al.*, 1988), both alternative pathways for NADPH production, suggested that this co-factor is probably not rate-limiting in DHEA-treated animals.

Another explanation for DHEA's antiobesity effects suggests induction of futile cycles. Mohan and Cleary (1988) proposed an ATP-consuming fatty acid acylation-deacylation cycle which was based on the observation that acyl-CoA hydrolase and acyl-CoA synthetase activities were both elevated after 3 and 7 days of DHEA-treatment. A second futile cycle involving enhanced glucose oxidation and reduced gluconeogenesis was suggested by McIntosh and Berdanier (1991).

Other metabolic pathways

Electron microscopic studies have shown that the number of mitochondria in hepatocytes is elevated in DHEA-treated rats (Metzger *et al.*, 1995; Bannasch *et al.*, 2000). This is reflected in the increase in content of mitochondrial protein (Mohan and Cleary, 1991) and activity of mitochondrial enzymes (Mayer *et al.*, 1998a). Glycerol-3-phosphate dehydrogenase (G3PDH), a mitochondrial flavoprotein which plays an important role in degradation of triglycerides and transport of NADH into mitochondria is strongly induced indicating a strong elevation of proton transport. Mitochondrial respiration was also found to be affected by DHEA-treatment. State 3 respiration with glutamate–malate and succinate as substrates was strongly elevated in obese Zucker rats after a 3 days-treatment and in lean rats after 7 days, as well as three hours after administration of a large single dose of DHEA by gastric intubation (100 mg DHEA/kg body weight) (Mohan and Cleary, 1988, 1991; Cleary, 1990). This points to an increase in electron flow through the mitochondrial respiratory chain, direct measurements of the efficiency of oxidative phosphorylation and ATP production have not been reported. Berdanier and Flatt (2000) suggest that the changes in thyroid status (McIntosh and Berdanier, 1991, 1992) probably result in an increase in the tightness of coupling respiration to ATP synthesis. The effects reported on mitochondrial activity might contribute to the overall energy wasting caused by DHEA.

Induction of peroxisome proliferation by DHEA in rat liver has been of specific importance in the context of the hepatocarcinogenic effect of DHEA (see below).

Increase in the number of peroxisomes has also a strong impact on the energy metabolism of liver. Peroxisome proliferation is accompanied by induction of acyl-CoA oxidase and peroxisomal hydratase/dehydrogenase, enzymes of peroxisomal β-oxidation. Ten to 40-fold induction of peroxisomal hydratase/dehydrogenase mRNA has been observed in rat liver as early as 24 hours after administration of a single large intragastric dose of DHEA (0.5 g/kg body weight) (Rao *et al.*, 1992a). Expression remained high after 6 weeks (Leighton *et al.*, 1987) and 24 weeks of treatment (Beier *et al.*, 1997). Other peroxisomal enzymes such as urate oxidase and catalase expession were found unchanged (Beier *et al.*, 1997). Induction of peroxisomal β-oxidation and peroxisome proliferation were suggested to result in an enhanced flux of free fatty acids through the peroxisomal β-oxidation pathway resulting in increased utilization of O_2 and elevated production of H_2O_2. Since peroxisomal β-oxidation, different from mitochondrial β-oxidation, does not result in ATP-production, but consumes ATP, this process contributes to the energy wasting in liver metabolism observed with DHEA. However, it is unlikely that peroxisome proliferation explains the antiobesity effect of DHEA, since other compounds that also increase peroxisome proliferation do not cause a weight loss.

Liver is the main site for *de novo* synthesis of cholesterol. Cholesterol is produced at high amounts as a precursor for the biosynthesis of bile acids, and it is required for cell growth and proliferation because cellular membranes are constituted to a large part of cholesterol. DHEA has been described to exert an inhibitory effect *in vitro* on the incorporation of radioactively labeled acetate into cholesterol in slices of normal liver gained from rats fed for 15 days with a diet containing 0.6% DHEA (Feo *et al.*, 1991). This effect could be abolished when acetate was substituted by mevalonate. It was concluded that DHEA inhibits a step of cholesterol biosynthesis which precedes mevalonate biosynthesis. The activity of 3-hydroxymethyl glutaryl-CoA reductase (HMGR), the first rate limiting enzyme in cholesterol biosynthesis was found significantly reduced in DHEA-treated rat liver which suggests that this enzyme is the target for DHEA inhibition (Pascale *et al.*, 1995). However, DHEA did not directly inhibit HMGR in an *in vitro* assay. Using a protocol of experimental hepatocarcinogenesis resulting in a large number of proliferating nodules (see below), Pascale *et al.* (1995) could show that DHEA decreases HMGR gene expression.

PEROXISOME PROLIFERATION

A number of researchers have shown induction of peroxisome proliferation in rat and mouse liver by dietary administration of DHEA (Leighton *et al.*, 1987; Frenkel *et al.*, 1990; Yamada *et al.*, 1991; Rao *et al.*, 1992a, 1994; Hayashi *et al.*, 1994; Prough *et al.*, 1994; Metzger *et al.*, 1995; Beier *et al.*, 1997). Requirement of sulfate conjugation for this effect (Yamada *et al.*, 1994; Ram and Waxman, 1994) suggests that DHEAS is mainly responsible for peroxisome proliferation. Increase in peroxisomal volume density of 2.5–11-fold has been observed both in male and female rats fed for two weeks a diet containing 0.45% DHEA (Rao *et al.*, 1994), this phenomenon showing no sex specificity. Nevertheless, other researchers have clearly documented a sex specificity with stronger effects in male (Yamada *et al.*, 1991) or in female rats (Leighton *et al.*, 1987; Metzger *et al.*, 1995; Beier *et al.*, 1997). Beier *et al.* (1997) described a strong gradient in volume density of peroxisomes in the liver acinus in female rats, hepatocytes surrounding the central terminal vein showing a very strong expression of peroxisomal hydratase/dehydrogenase after 78 weeks of treatment with 0.6% DHEA in the diet. Distribution of peroxisome proliferation and expression of peroxisomal hydratase was much more homogenous in the liver of male rats.

Concomitant with peroxisome proliferation and induction of enzymes of peroxisomal β-oxidation, a significant increase in the expression of cytochrome P450 IVA and NADPH-cytochrome P450 reductase was observed (Wu *et al.*, 1989; Prough *et al.*, 1994; Beier *et al.*, 1997). Both activities are accompanied by the production of reactive oxygen species, and it has been suggested that increased peroxisomal H_2O_2-production is causally related to induction of hepatocellular neoplasms. Catalase, the peroxisomal H_2O_2-degrading enzyme, was found to be unchanged (Beier *et al.*, 1997) or only weakly induced (Mayer *et al.*, 1996) in DHEA-treated livers. Catalase activity, however, was quite high in untreated liver (Hayashi *et al.*, 1994; Metzger *et al.*, 1995; Mayer *et al.*, 1996, 1998a; Beier *et al.*, 1997), and it is unclear whether there is any significant local increase at all in H_2O_2-concentration in DHEA-treated liver.

The lobular distribution of the microsomal cytochrome P450 IVA, which was strongly induced in males and only moderately increased in females by DHEA, showed a striking homology with peroxisomal enzymes in both sexes (Beier *et al.*, 1997), indicating a functional relationship. It may be speculated that the strong induction of cytochrome P450 IVA in males is related to the smaller incidence of DHEA-induced neoplasms due to accelerated metabolism of the substance. This assumption is confirmed by the rapid disappearance of DHEA from the blood of male rats observed by Hobe *et al.* (1994a,b).

The mechanisms underlying induction of peroxisome proliferation by DHEA remain unclear. For many peroxisome proliferating agents it has been shown that both the significant increase in cytochrome P450 IVA and the induction of the enzymes of the peroxisomal β-oxidation are mediated by the peroxisome proliferator activated receptors (PPARs) (Issemann and Green, 1990). Although DHEA does not activate the PPAR in *in vitro* assays (Issemann and Green, 1990; Göttlicher *et al.*, 1992; Waxman, 1996), more recent studies have revealed that the peroxisome induction by DHEA is completely abolished in knock-out mice with deletion of the PPAR-α gene (Peters *et al.*, 1996), which indicates the involvement of this receptor in the induction process.

There is evidence from studies on cultured rat hepatocytes that activation of Ca^{2+}-dependent signaling pathways is required for induction of peroxisome proliferation by DHEA. The Ca^{2+}-antagonist nicardipine abolished induction of peroxisomal β-oxidation by DHEAS (Zhang *et al.*, 1996). It remains to be noted that mice are less responsive to peroxisome proliferation than rats; rainbow trout, hamster, guinea pig and primates including humans (cultured hepatocytes) show no or only very little susceptibility to peroxisome proliferation.

OXIDATIVE STRESS AND LIPID PEROXIDATION

Increased peroxisomal, microsomal and mitochondrial activity induced by DHEA in rat liver may result in the production of reactive oxygen species and, consequently, liver injury. DHEA-administration to Sprague-Dawley rats caused a dose-dependent increase in NADPH-dependent lipid peroxidation in microsomes and mitochondria isolated from rat liver of both genders (Swierczynski *et al.*, 1996; Swierczynski and Mayer, 1996). Significant increase in lipid peroxidation was observed as early as two and three days after start of the treatment with a maximum after seven and 14 days in males and females, respectively. The effect could be prevented by administration of vitamin E (Swierczynski *et al.*, 1997). Males were more susceptible to lipid peroxidation than females, which agreed with the strong induction of NADPH-cytochrome P 450 reductase/cytochrome P450, an H_2O_2-producing enzyme system, by DHEA, in the livers of male rats (Wu *et al.*, 1989; Prough *et al.*, 1994; Beier *et al.*, 1997). Furthermore, mitochondrial proliferation induced by DHEA (Metzger *et al.*, 1995) may contribute to the increase in lipid peroxidation, since reactive oxygen species are regularly produced by electron leakage in the mitochondrial respiratory chain, particularly by complex III. It remains to be clarified if the products of lipid peroxidation, e.g. malondialdehyde (Swierczynski *et al.*, 1996), reacts with DNA leading to DNA damage.

HEPATOCARCINOGENIC EFFECT OF DHEA

A hepatocarcinogenic effect of DHEA has been observed in rats and rainbow trout independently in several laboratories (Rao *et al.*, 1992b,c; Hayashi *et al.*, 1994; Metzger *et al.*, 1995; Orner *et al.*, 1995).

In two different strains of rat, long-term administration of high doses of DHEA in the diet resulted in well-differentiated trabecular or solid hepatocellular neoplasms (Table 5.1) (Metzger *et al.*, 1995). Rao and colleagues (1992c) found hepatocellular carcinomas in 88% of male F-344 rats exposed to 0.45% (w/w) DHEA in the diet for 70–84 weeks. After feeding DHEA to male F-344 rats at two different dose levels (0.5% and 1%) for 52 and 72 weeks, Hayashi *et al.* (1994) have shown that the hepatocarcinogenic effect of DHEA is dose- and time-dependent, resulting in a tumor incidence of 100% after exposure to the higher dose level for 72 weeks. Metzger *et al.* (1995) studied the effect of dietary DHEA (0.6%) in both male and female Sprague-Dawley rats. Hepatocellular adenomas developed in 33% of the female rats after 70–84 weeks of treatment, but only rarely in male rats. Carcinomas were found in 44% of the females and in 11% of the males (Table 5.1), but metastases were never observed. Rao and Subbarao (1998) observed more adenomas and carcinomas in male than in female F-344 rats. In these experiments male rats developed mainly carcinomas, while females developed mainly adenomas. Ovariectomy increased the number of total tumors, but did not significantly increase the incidence of hepatocellular carcinomas, a finding which could not be explained (Rao and Subbarao, 1998). It was discussed by Labrie and co-workers (1989) and Sourla and co-workers (1998) that DHEA exerts a mainly androgenic effect on both male and female rats. This suggests a different sensitivity of the two rat strains studied to the carcinogenic effect of androgenic hormones. The lower tumor incidence observed in Sprague-Dawley rats (Metzger *et al.*, 1995) as compared to the findings reported by Rao and colleagues (1992c) and Hayashi and coworkers (1994) may be due to the generally lower susceptibility of Sprague-Dawley rats,

Table 5.1 Incidence of hepatocellular neoplasms induced by dietary DHEA in different rat strains

Rat strain	Sex	Dose	Duration of treatment	Total incidence of hepatocellular neoplasms	HCC incidence	Reference
F-344	male	0.45%	70–84 wk	94%	88%	Rao et al., 1992c
F-344	male	0.45%	100 wk	94%	81%	Rao and Subbarao, 1998
	female	0.45%	100 wk	46%	15%	
	female (ovariect.)	0.45%	100 wk	60%	20%	
F-344	male	0.5%	52 wk	20%	n.d.[a]	Hayashi et al., 1994
		0.5%	72 wk	56%		
		1.0%	52 wk	20%		
		1.0%	72 wk	100%		
Sprague-Dawley	male	0.6%	70–84 wk	16%	11%	Metzger et al., 1995
	female	0.6%	70–84 wk	55%	44%	

Note
[a]n.d., not described.

compared to F-344 rats, to hepatocarcinogenic agents. In Sprague-Dawley rats, the higher tumor incidence in females correlates with the higher DHEA(S) levels measured in the plasma as compared to male rats (Hobe *et al.*, 1994a).

As to the significance of lipid peroxidation for induction of hepatocellular neoplasms it is of importance to note that the increase in lipid peroxidation is an early transient effect of DHEA (Swierzcynski and Mayer, 1996; Swierczynski *et al.*, 1996, 1997), but liver neoplasms occur only after long-term administration. Thus, lipid peroxidation seems to represent rather an unspecific process leading to liver injury which may, nevertheless, be involved in hepatocarcinogenesis by eliciting cell proliferation.

DHEA proved to be a complete carcinogen in rainbow trout. Feeding of DHEA produced liver tumors, most of which were classified as mixed hepatocholangiocellular carcinomas (60%), and hepatocellular carcinomas (30%) (Orner *et al.*, 1995). The remainder consisted of mixed adenomas, hepatocellular adenomas and cholangiomas. The neoplasms developed at dose levels as low as 444–888 p.p.m. In contrast to rat liver where no mutations induced by DHEA were described so far, Ki-*ras* mutations (transitions in codon 12 and transversions in codon 13) were observed in 20–41% of the DHEA-induced liver neoplasms in trout. These observations are of particular interest for the understanding of the mechanism of DHEA-induced hepatocarcinogenesis, since rainbow trout, like humans, are weak responders to peroxisome proliferators.

The amphophilic cell lineage of hepatocellular neoplasms

Hepatocellular neoplasms (adenomas and carcinomas) induced by DHEA in the rat develop from small preneoplastic lesions or foci of altered hepatocytes which were classified as amphophilic cell foci (APF) (Weber *et al.*, 1988a). Light microscopically the phenotype of APFs is characterized by a homogenous cytoplasmic acidophilia

Table 5.2 Enzyme alterations of glycogen storing foci, amphophilic cell foci, basophilic cell foci and hepatocellular adenomas and carcinomas as compared to normal liver tissue[a] (adapted from Mayer et al., 1998a)

Parameter	Type of lesion							
	GSF		APF		HCA + MC, amphophilic		HCC	
	n = 155	%	n = 84	%	n = 42	%	n = 8	%
Glycogen	●●●	100	○○○	100	○○○	93	○○○	88
Basophilia	○	60	●●●	100	●●●	100	●●●	100
SYN	(●)	49	○○	75	○○	72	○○○	100
PHO	○	58	○○	78	○○○	85	○○	80
HK	n.c.	76	n.c.	90	n.c.	70	●	57
PK	(●)	43	○○	74	○○	64	○○	75
G6PDH	●●●	96	n.c.	64	●	66	●●●	100
ME	●●●	94	n.c.	64	●	53	●●●	88
G6Pase	○○	76	●	60	●	55	●	63
COX	n.c.	77	●●	75	●●	78	●●●	86
G3PDH	n.c.	64	●●	79	●●●	98	●●	71
SDH	n.c.	65	(●)	43	●	65	●	63
GIDH	●●	70	○	64	○○	77	○	50
AcPase	○	67	●●	76	○○	77	●	63
GGT	●●●	90	n.c.	91	n.c.	100	n.c.	57
PH	○○○	90	●●	75	○	50	○/●	50/50
Catalase	○○○	100	n.c.	50	n.c.	83	n.c.	100
AOX	○○○	90	●	70	○	50	●●	75

Notes

[a]The table shows the prevailing groups of the respective lesions which have either increased, decreased or unchanged activity, and the percentage of lesions which show the respective activity changes. ●●●, increase in >80%; ●●, increase in 70–80%; ●, increase in 50–70%; (●), increase in <50% of the lesions. ○○○, decrease in >80%; ○○, decrease in 70–80%; ○, decrease in 50–70% of the lesions; n.c., no change; n.d., not determined. Abbreviations: GSF, glycogen storing foci; APF, amphophilic cell foci; BCF, basophilic cell foci; HCA, hepatocellular adenomas; HCC, hepatocellular carcinomas; MC, microcarcinomas; PH, peroxisomal hydratase, AOX, acyl-CoA oxidase.

associated with a distinct diffuse or somewhat scattered basophilia. The fine structure of amphophilic cells shows a strikingly high content of cristae-rich mitochondria wrapped in cisternae of the rough endoplasmic reticulum, and a slightly increased number of peroxisomes. During progression to larger focal lesions and neoplasms, the cellular phenotype often changes from the amphophilic to an amphophilic/tigroid character with an increased structured basophilia (Weber et al., 1988a; Metzger et al., 1995). This amphophilic cell lineage of hepatocarcinogenesis is characteristic for DHEA and probably other peroxisome proliferators and as a rule does not occur with genotoxic hepatocarcinogens such as nitrosamines. Amphophilic cell foci and hepatocellular neoplasms arising from these lesions show a unique biochemical phenotype (Table 5.2). They are characterized by a reduction in glycogen content and activity of glycogen metabolizing enzymes, increase in glucose-6-phosphatase activity, a key enzyme in gluconeogenesis, increase in mitochondrial and peroxisomal enzymes, and increase in lysosomal acid phosphatase. It was suggested that this pattern of enzyme expression mimics the effect of thyroid hormones. Therefore, DHEA was attributed a thyromimetic effect on hepatocytes in the rat (Bannasch *et al.*, 1997; Mayer *et al.*, 1998a).

In contrast to the predominantly perivenular peroxisome proliferation, the amphophilic cell populations emerged from cells localized in periportal and intermediate parts of the liver lobules, largely corresponding to zone 1 of the functional liver acinus as defined by Rappaport. It is, therefore, questionable, if peroxisome proliferation is causally related to hepatocarcinogenesis induced by DHEA.

Enhancement of hepatocarcinogenesis by DHEA

DHEA may act as a tumor promoter in rat and trout liver. It enhances hepatocarcinogenesis induced by the strong (genotoxic) hepatocarcinogen *N*-nitroso-morpholine (NNM) in rats when given subsequent to the nitrosamine for up to 32 weeks (Metzger *et al.*, 1997). While the combined tumor incidence (adenomas and carcinomas) was similar in male (63%) and female (60%) rats after NNM-treatment alone, tumor incidence was higher in females (87%) than in males (80%) after combined NNM/DHEA-treatment. The difference between the genders was mainly due to the higher incidence of hepatocellular carcinoma in females. Orner and co-workers (1995, 1998) observed a strong dose and time-dependent tumor promoting activity of DHEA in trout pre-treated with aflatoxin B1. From these data it is evident that DHEA is a full hepatocarcinogen, and an enhancer of hepatocarcinogenesis induced by other agents in rats and trout. While in trout the two genders were not investigated separately, female Sprague-Dawley rats proved to be more sensitive to the tumor enhancing activity of DHEA than males.

MODULATION OF HEPATOCARCINOGENESIS BY DHEA

DHEA-treatment not only results in the induction of APF, but also causes a modulation of glycogen storage foci (GSF) typically induced in rat liver by NNM into APF (Bannasch *et al.*, 1980; Weber *et al.*, 1988b; Metzger *et al.*, 1997). The modulation of the morphological phenotype is accompanied by a far-reaching modulation of the biochemical phenotype (Mayer *et al.*, 1998a). The enzyme pattern of APF differs significantly from that observed in focal lesions of the glycogenotic/basophilic cell lineage induced by NNM alone and indicates a fundamental shift in energy metabolism induced by DHEA (Mayer *et al.*, 1998a). Table 5.2 summarizes the patterns of enzymatic aberrations in preneoplastic and neoplastic lesions of the glycogenotic/basophilic and the amphophilic cellular lineages. A biplot analysis reveals that the alterations in enzyme activities observed in GSF and in APF are in many respects opposite (Figure 5.1). This applies particularly to G6PDH and malic enzyme which are strongly increased in GSF but unchanged in APF, and to gluconeogenic G6Pase which is reduced in GSF but increased in APF. Mitochondrial enzymes are increased in APF but unchanged in GSF, and the peroxisomal enzymes are decreased in GSF and increased (except catalase) in APF. The amphophilic neoplasms (hepatocellular adenomas and microcarcinomas) obtained by NNM/DHEA-treatment showed a similar pattern as the APF.

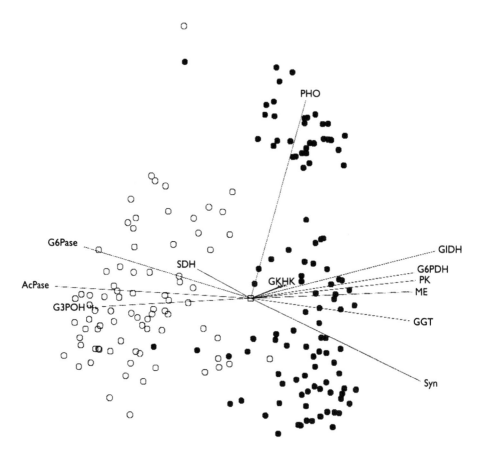

Figure 5.1 Biplot analysis of enzyme alterations in glycogen storage foci, GSF, (●) and amphophilic cell foci, APF, (○). Each symbol represents an individual focus. The lines from the origin show the positive direction of the changes of enzyme activities, and their length explains the importance of the enzymes for the distinction of the two types of lesions. Enzymes localized around the horizontal axis of the biplot represent markers for the populations investigated. It is evident that GSF and APF represent two populations with different enzyme patterns, G3PDH and G6PDH being possible markers for amphophilic foci and glycogen storing foci, respectively (from Mayer et al., 1998a).

The two patterns of enzyme aberrations are striking in the way that in the glyco-genotic/basophilic cell lineage the metabolic situation seems to be energy preserving, whereas that in the amphophilic lineage resembles an energy wasting state. It has been suggested previously that the pattern of enzyme alterations in GSF mimics an insulin effect (insulinomimetic effect of NNM) which contrasts the pattern observed in APF that resembles a thyroid hormone response (thyromimetic effect of DHEA) (Bannasch *et al.*, 1997; Mayer *et al.*, 1998a). The energy wasting thyromimetic state in APF and related neoplasms may explain the reduced mitotic index suggesting a reduction in growth rate (Weber *et al.*, 1988b). However, it cannot explain the tumor promoting effect of DHEA (Metzger *et al.*, 1997).

Hormone-like effects have been described for a large variety of chemical carcinogens. DHEA is metabolized to compounds with estrogenic and androgenic activity in rat liver (Hobe *et al.*, 1994a,b, 2000). It has also been shown that induction of peroxisome proliferation by DHEA is dependent on the expression of the peroxisome proliferator activated receptor-α (Peters *et al.*, 1996), a member of the steroid receptor superfamily. Since the thyroxine plasma level was reported to be slightly increased in DHEA-treated rats (McIntosh, 2000) and the thyroid hormone receptor is also a member of the steroid receptor family, it is tempting to speculate that a cross-talk between the receptor subtypes may be involved somehow in the thyromimetic response of preneoplastic hepatocytes to DHEA.

ANTICARCINOGENIC EFFECT OF DHEA

DHEA has been observed to show a certain anti-cancer effect in various organs in rodents subjected to chemically induced carcinogenesis (reviewed in Mayer, 1998; Feo *et al.*, 2000). As to the liver the situation is very complex since both cancer enhancing and anti-cancer activities of DHEA have been reported depending on the experimental protocol used and the end-points of the carcinogenic response studied (Table 5.3). Many authors use the expression of γ-glutamyltranspeptidase (GGT) and glutathione-*S*-transferase placental form (GSTP) as markers for the detection of preneoplastic liver lesions. These enzymes, however, are not expressed in APF and hepatocellular neoplasms induced by DHEA and in lesions induced by NNM/DHEA-treatment. Hence, the interpretation of an "anticarcinogenic" effect of DHEA requires some prudence. Nevertheless, a certain anticarcinogenic effect of DHEA has been described by a number of authors using histological and morphological markers (Table 5.3).

The inhibitory effect of DHEA on hepatocarcinogenesis was usually attributed to an inhibition of G6PDH activity by DHEA and to a reduction of NADPH. However,

Table 5.3 Anticarcinogenic effect of dehydroepiandrosterone in rat liver. Schedule of administration of dehydroepiandrosterone and of the carcinogens are specified in the references

Rat strain	Protocol	Lesion	Identification	Reference
Wistar	initiation/selection	foci and nodules	GGT/G6PDH	Garcea *et al.*, 1987
Wistar	initiation/selection	foci	GGT	Feo *et al.*, 1991
Fischer 344	initiation/promotion	nodules, carcinoma	G6PDH, histology	Simile *et al.*, 1995
Fischer 344	DHPN	preneoplastic lesions	GSF, GST-P	Moore *et al.*, 1986
Fischer 344	Azaserine	foci	GST-P	Thornton *et al.*, 1989
Fischer 344	DAB	MCF/BCF/APF	histology	Moore *et al.*, 1988b
Sprague-Dawley	NNM	angiosarcoma	histology	Weber *et al.*, 1988b

Notes
GGT, γ-glutamyltransferase; G6PDH, glucose-6-phosphate dehydrogenase; GSF, glycogen storage focus; GST-P, glutathione S-transferase placental form; DAB, dimethylaminoazobenzene; NNM, N-nitrosomorpholine; MCF, mixed cell focus; BCF, basophilic cell focus; APF, amphophilic cell focus.

as outlined above, NADPH seems not to be the limiting factor for cell growth since other NADPH producing enzymes such as malic enzyme and isocitrate dehydrogenase are induced in DHEA-treated liver (Garcea *et al.*, 1988; Bobyleva *et al.*, 1993; Mayer *et al.*, 1996). Using an initiation/selection model of hepatocarcinogenesis which combines partial hepatectomy and treatment with several chemical carcinogens and results in a large number of rapidly proliferating nodules, Garcea *et al.* (1987) observed a reduction of G6PDH activity in total liver postmitochondrial supernatant and a reduction in the number of GGT-positive liver lesions by DHEA-treatment. The same authors (Garcea *et al.*, 1988; Feo *et al.*, 1991) reported an inhibition of [³H]-thymidine incorporation and a reduction of mitotic index in rapidly proliferating hepatic nodules induced in female Wistar rats. The inhibition of thymidine incorporation could be reversed by the injection of ribo- and deoxyribonucleosides (Garcea *et al.*, 1988). The authors conclude that DHEA inhibits formation of preneoplastic lesions by the reduction of G6PDH activity resulting in a depletion of the tissue in ribo- and deoxyribonucleotides. Weber *et al.* (1988b) treated male Sprague-Dawley rats with *N*-nitrosomorpholine (NNM) at 120 mg/litre drinking water for 7 weeks, and with DHEA (0.25% [w/w] in the diet) throughout the experiment. They observed a significant reduction of the number of hepatic angiosarcomas by DHEA-treatment. Furthermore, a modulation of the cellular phenotype but no significant influence on the overal incidence of hepatocellular neoplasms was observed. The total number of hepatocellular carcinomas did not change significantly, whereas the incidence of poorly differentiated carcinomas decreased slightly in favour of highly differentiated hepatocellular carcinomas. In addition, DHEA-treatment resulted in a reduction of the mitotic index of the tumors. Thus, under conditions where NNM- and DHEA-treatment were started simultaneously, DHEA had a certain anticarcinogenic effect in the liver. A modulation of preneoplastic lesions from the glycogenotic to the amphophilic phenotype by DHEA was also observed when male Sprague-Dawley rats were treated with dimethylaminoazobenzene, followed by DHEA (Moore *et al.*, 1988a). The number and area of mixed cell populations was decreased, but instead an increase in APF occurred.

The data reported above indicate that DHEA does exert a certain anticarcinogenic effect in rat liver. DHEA reduces growth rate of preneoplastic and neoplastic liver lesions and induces a shift of their phenotype towards less malignant forms. However, the experimental conditions and schedule of administration of the substances investigated as well as the methods used for detecting the anticarcinogenic effects are crucial and must be taken into consideration in order to avoid misinterpretation.

CONCLUSION

DHEA exerts multiple effects on rat liver which are similar in males and females. In both genders DHEA induces alterations in the expression pattern of key enzymes of energy metabolism, mainly lipid and carbohydrate metabolism, which indicate profound changes in liver function. The alterations result in an overall waste of energy which may partly explain the anti-obesity and anti-cancer effects of DHEA. However, DHEA also acts as a hepatocarcinogen and tumor promoter in rats, females being more sensitive than males. The molecular and biochemical mechanisms related to the DHEA-induced hepatocarcinogenesis require further investigation.

REFERENCES

Bannasch, P., Klimek, F. and Mayer, D. (1997) Early bioenergetic changes in hepatocarcinogenesis: preneoplastic phenotypes mimic responses to insulin and thyroid hormone. *J. Bioenerg. Biomemb.*, **29**, 303–313.

Bannasch, P., Mayer, D. and Hacker, H.J. (1980) Hepatocellular glycogenosis and hepatocarcinogenesis. *Biochim. Biophys. Acta*, **605**, 217–245.

Bannasch, P., Metzger, C. and Mayer, D. (2000) Hepatocarcinogenesis by dehydroepiandrosterone. I. Induction of neoplasms and sequential cellular changes during neoplastic development. In M. Kalimi and W. Regelson (eds), *Dehydroepiandrosterone (DHEA), biochemical, physiological and clinical aspects*, Walter de Gruyter, Berlin, pp. 237–249.

Baulieu, E.E. (1962) Studies of conjugated 17-ketosteroids in a case of adrenal tumor. *J. Clin. Endocrinol. Metab.*, **22**, 501–510.

Baulieu, E.E. and Robel, P. (1996) Dehydroepiandrosterone and dehydroepiandrosterone sulfate as neuroactive neurosteroids. *J. Endocrinol.*, **150**, S221–S239.

Beier, K., Völkl, A., Metzger, C., Mayer, D., Bannasch, P. and Fahimi, H.D. (1997) Hepatic zonation of the induction of cytochrome P450 IVA, peroxisomal lipid β-oxidation enzymes and peroxisome proliferation in rats treated with dehydroepiandrosterone (DHEA). Evidence of distinct sex differences. *Carcinogenesis*, **18**, 1491–1498.

Bellino, F.L., Daynes, R.A., Hornsby, P.J., Lavrin, D.H. and Nestler, J.E. (eds) (1995) Dehydroepiandrosterone and aging. *Ann. N. Y. Acad. Sci.*, **774**, 1–350.

Berdanier, C.D. and Flatt, W.P. (2000) DHEA and mitochondrial respiration. In M. Kalimi and W. Regelson (eds), *Dehydroepiandrosterone (DHEA), biochemical, physiological and clinical aspects*, Walter de Gruyter, Berlin, pp. 377–391.

Bobyleva, V., Kneer, N., Bellei, M., Battelli, D. and Lardy, H.A. (1993) Concerning the mechanism of increased thermogenesis in rats treated with dehydroepiandrosterone. *J. Bioenerg. Biomemb.*, **25**, 313–321.

Bradlow, H.L. and Zumoff, B. (2000) Androgens and liver function. In M. Kalimi and W. Regelson (eds), *Dehydroepiandrosterone (DHEA), biochemical, physiological and clinical aspects*, Walter de Gruyter, Berlin, pp. 363–375.

Casazza, J.P., Schaffer, T. and Veech, R.L. (1986) The effect of dehydroepiandrosterone on liver metabolites. *J. Nutr.*, **116**, 304–310.

Cleary, M.P. (1990) Effect of dehydroepiandrosterone treatment on liver metabolism in rats. *Int. J. Biochem.*, **22**, 205–210.

Cleary, M.P. (2000) The antiobesity effect of dehydroepiandrosterone treatment. In M. Kalimi and W. Regelson (eds), *Dehydroepiandrosterone (DHEA), biochemical, physiological and clinical aspects*, Walter de Gruyter, Berlin, pp. 157–172.

Cleary, M.P., Fox, N., Lazin, B. and Billheimer, J.T. (1985) A comparison of the effects of dehydroepiandrosterone treatment to ad libitum and pair-feeding in the obese Zucker rats. *Nutr. Res.*, **5**, 1247–1257.

Cleary, M.P., Hood, S.S., Chando, C., Hansen, C.T. and Billheimer, J.T. (1984a) Response of sucrose-fed BHE-rats to dehydroepiandrosterone. *Nutr. Res.*, **4**, 485–494.

Cleary, M.P., Shepherd, A. and Jenks, B. (1984b) Effect of dehydroepiandrosterone on growth in lean and obese Zucker rats. *J. Nutr.*, **114**, 1242–1251.

Cleary, M.P., Shepherd, A., Zisk, J. and Schwartz, A. (1983) Effect of dehydroepiandrosterone on body weight and food intake in rats. *Nutrition and Behavior*, **1**, 127–136.

Diersen-Schade, D.A. and Cleary, M.P. (1989) No effects of long-term DHEA on either hepatocyte and adipocyte pentose pathway activity or adipocyte glycerol release. *Horm. Metab. Res.*, **21**, 356–358.

Doostzadeh, J., Cotillon, A.C., Benalychérif, A. and Morfin, R. (1998) Inhibition studies of dehydroepiandrosterone 7α- and 7β-hydroxylation in mouse liver microsomes. *Steroids*, **63**, 608–614.

Feo, F., Daino, L., Seddaiu, M.A., Simile, M.M., Pascale, R.M., McKeating, J.A. *et al.* (1991) Differential effects of dehydroepiandrosterone and deoxyribonucleotides on DNA synthesis and *de novo* cholesterogenesis in hepatocarcinogenesis in rat. *Carcinogenesis*, **12**, 1581–1586.

Feo, F., Pascale, R.M., Simile, M.M. and De Miglio, M.R. (2000) Role of dehydroepiandrosterone in experimental and human carcinogenesis. In M. Kalimi and W. Regelson (eds), *Dehydroepiandrosterone (DHEA), biochemical, physiological and clinical aspects*, Walter de Gruyter, Berlin, pp. 215–236.

Finan, A.C. and Cleary, M.P. (1988) Lack of an effect of short-term dehydroepiandrosterone treatment on the pentose pathway. *Nutr. Res.*, **8**, 755–765.

Frenkel, R.A., Slaughter, C.A., Orth, K., Moomaw, C.R., Hicks, S.H., Snyder, J.M. *et al.* (1990) Peroxisome proliferation and induction of peroxisomal enzymes in mouse and rat liver by dehydroepiandrosterone feeding. *J. Steroid Biochem.*, **35**, 333–342.

Garcea, R., Daino, L., Frassetto, S., Cozzolino, P., Ruggiu, M.E., Vannini, M.G. *et al.* (1988) Reversal by ribo- and deoxyribonucleosides of dehydroepiandrosterone-induced inhibition of enzyme altered foci in the liver of rats subjected to the initiation-selection process of experimental hepatocarcinogenesis. *Carcinogenesis*, **9**, 931–938.

Garcea, R., Daino, L., Pascale, R., Frassetto, S., Cozzolino, P., Ruggiu, M.E. *et al.* (1987) Inhibition by dehydroepiandrosterone of liver preneoplastic foci formation in rats after initiation-selection in experimental carcinogenesis. *Toxicol. Pathol.*, **15**, 164–169.

Göttlicher, M., Widmark, E., Li, Q. and Gustafsson, J.A. (1992) Fatty acids activate a chimera of the clofibric acid-activated receptor and the glucocorticoid receptor. *Proc. Natl. Acad. Sci. USA*, **89**, 4653–4657.

Hamilton, S.R., Gordon, G.B., Floyd, J. and Golightly, S. (1991) Evaluation of dietary dehydroepiandrosterone for chemoprotection against tumorigenesis in premalignant colonic epithelium of male F-344 rats. *Cancer Res.*, **51**, 476–480.

Hayashi, F., Tamura, H., Yamada, J., Kasai, H. and Suga, T. (1994) Characteristics of the hepatocarcinogenesis caused by dehydroepiandrosterone, a peroxisome proliferator, in male F-344 rats. *Carcinogenesis*, **15**, 2215–2219.

Hobe, G., Hillesheim, H.G., Schön, R., Reddersen, G., Knappe, R., Bannasch, P. *et al.* (1994a) Sex differences in dehydroepiandrosterone metabolism in the rat: different plasma levels following ingestion of DHEA-supplemented diet and different metabolite patterns in plasma, bile and urine. *Horm. Metab. Res.*, **26**, 326–329.

Hobe, G., Hillesheim, H.-G., Schön, R., Ritter, P., Reddersen, G., Mayer, D. *et al.* (1994b) Sex differences in dehydroepiandrosterone (DHEA) metabolism in the rat. In S. Görög (ed.), *Advances in steroid analysis '93*. Akadémiai Kiádó, Budapest, pp. 383–390.

Hobe, G., Hillesheim, H.-G., Schön, R., Undisz, K., Valentin, U., Reddersen, G. *et al.* (2000) Studies on metabolism of DHEA in rats and mice. In M. Kalimi and W. Regelson (eds), *Dehydroepiandrosterone (DHEA), biochemical, physiological and clinical aspects*, Walter de Gruyter, Berlin, pp. 343–362.

Issemann, I. and Green, S. (1990) Activation of a member of the steroid hormone receptor superfamily by peroxisome proliferators. *Nature*, **347**, 645–650.

Kalimi, M. and Regelson, W. (eds) (1990) *The biologic role of dehydroepiandrosterone (DHEA)*, Walter de Gruyter, Berlin, New York.

Kalimi, M. and Regelson, W. (eds) (2000) *Dehydroepiandrosterone (DHEA)*, Walter de Gruyter, Berlin, New York.

Kullak-Ublick, G.A., Fisch, T., Oswald, M., Hagenbuch, B., Meier, P.J., Beuers, U. *etal.* (1998) Dehydroepiandrosterone sulfate (DHEAS): identification of a carrier protein in human liver and brain. *FEBS Lett.*, **424**, 173–176.

Labrie, C., Simard, J., Zhao, H.F., Bélanger, A., Pelletier, G. and Labrie, F. (1989) Stimulation of androgen-dependent gene expression by the adrenal precursors dehydroepiandrosterone and androstenedione in the rat ventral prostate. *Endocrinology*, **124**, 2745–2754.

Lardy, H., Kneer, N., Wie, Y., Partridge, B. and Marwah, P. (1998) Ergosteroids. II. Biologically active metabolites and synthetic derivatives of dehydroepiandrosterone. *Steroids*, **63**, 158–165.

Leighton, B., Tagliaferro, A.R. and Newsholme, E.A. (1987) The effect of dehydroepiandrosterone acetate on liver peroxisomal enzyme activities of male and female rats. *J. Nutr.*, **117**, 1287–1290.

Marks, P.A. and Banks, J. (1960) Inhibition of mammalian glucose-6-phosphate dehydrogenase by steroids. *Proc. Natl. Acad. Sci.*, **45**, 447–452.

Mayer, D. (1998) Carcinogenic and anticarcinogenic effects of dehydroepiandrosterone in the liver of male and female rats. *The Aging Male*, **1**, 56–66.

Mayer, D., Metzger, C. and Bannasch, P. (2000) Hepatocarcinogenesis by dehydroepiandrosterone. II. Biochemical and molecular changes during neoplastic development. In M. Kalimi and W. Regelson (eds), *Dehydroepiandrosterone (DHEA), biochemical, physiological and clinical aspects*, Walter de Gruyter, Berlin, pp. 251–260.

Mayer, D., Metzger, C., Leonetti, P., Beier, K., Benner, A. and Bannasch, P. (1998a) Differential expression of key enzymes of energy metabolism in preneoplastic and neoplastic rat liver lesions induced by *N*-nitrosomorpholine and dehydroepiandrosterone. *Int. J. Cancer Pred. Oncol.*, **79**, 232–240.

Mayer, D., Metzger, C., Nehrbass, D. and Bannasch, P. (1998b) Hepatocarcinogenesis by dehydroepiandrosterone. *Current Topics in Steroid Research*, **1**, 135–144.

Mayer, D., Reuter, S., Hoffmann, H., Bocker, T. and Bannasch, P. (1996) Dehydroepiandrosterone reduces expression of glycolytic and gluconeogenic enzymes in the liver of male and female rats. *Int. J. Oncol.*, **8**, 1069–1078.

Mayer, D., Weber, E., Moore, M.A., Letsch, I., Filsinger, E. and Bannasch, P. (1988) Dehydroepiandrosterone (DHEA) induced alterations in rat liver carbohydrate metabolism. *Carcinogenesis*, **11**, 2039–2049.

McIntosh, M.K. (2000) DHEA(S) and obesity: potential antiadipogenic mechanisms of action. In M. Kalimi and W. Regelson (eds), *Dehydroepiandrosterone (DHEA), biochemical, physiological and clinical aspects*, Walter de Gruyter, Berlin, pp. 131–156.

McIntosh, M.K. and Berdanier, C. (1991) Antiobesity effects of DHEA are mediated by futile substrate cycling in the hepatocytes of BHE/cdb rats. *J. Nutr.*, **121**, 2037–2043.

McIntosh, M.K. and Berdanier, C. (1992) Influence of DHEA on the thyroid hormone status of BHE/cdb rats. *J. Nutr. Biochem.*, **3**, 194–199.

Metzger, C., Bannasch, P. and Mayer, D. (1997) Enhancement and phenotypic modulation of *N*-nitrosomorpholine-induced hepatocarcinogenesis by dehydroepiandrosterone. *Cancer Lett.*, **121**, 125–131.

Metzger, C., Mayer, D., Hobe, G., Hoffmann, H., Bocker, T., Benner, A. *etal.* (1995) Sequential appearance and ultrastructure of amphophilic cell foci, adenomas, and carcinomas in the liver of male and female rats treated with dehydroepiandrosterone. *Toxicol. Pathol.*, **23**, 591–605.

Mohan, P.F. and Cleary, M.P. (1988) Effect of short-term DHEA administration on liver metabolism of lean and obese rats. *Am. J. Physiol.*, **255**, E1–E8.

Mohan, P.F. and Cleary, M.P. (1991) Short-term effects of dehydroepiandrosterone treatment in rats on mitochondrial respiration. *J. Nutr.*, **121**, 240–250.

Moore, M.A., Thamavit, W., Tsuda, H., and Ito, N. (1986) The influence of subsequent dehydroepiandrosterone, diaminopropane, phenobarbital, butylated hydroxyanisole and butylated hydroxytoluene treatment on the development of preneoplastic and neoplastic lesions in the rat initiated with di-hydroxy-di-propyl nitrosamine. *Cancer Lett.*, **30**, 153–160.

Moore, M.A., Weber, E. and Bannasch, P. (1988a) Modulating influence of dehydro-epiandrosterone administration on the morphology and enzyme phenotype of dimethyl-aminoazobenzene-induced hepatocellular foci and nodules. *Virchows Arch. B Cell Pathol.*, **55**, 337–343.

Moore, M.A., Weber, E., Thornton, M. and Bannasch, P. (1988b) Sex-dependent, tissue-specific opposing effects of dehydroepiandrosterone on initiation and modulation stages of liver and lung carcinogenesis induced by dihydroxy-di-*N*-propylnitrosamine in F344 rats. *Carcinogenesis*, **9**, 1507–1509.

Orner, G.A., Hendricks, J.D., Arbogast, D. and Williams, D.E. (1998) Modulation of aflatoxin-B1 hepatocarcinogenesis in trout by dehydroepiandrosterone: initiation/post-initiation and latency effects. *Carcinogenesis*, **19**, 161–167.

Orner, G.A., Mathews, C., Hendricks, J.D., Carpenter, H.M., Bailey, G.S. and Williams, D.E. (1995) Dehydroepiandrosterone is a complete hepatocarcinogen and potent tumor promoter in the absence of peroxisome proliferation in rainbow trout. *Carcinogenesis*, **16**, 2893–2898.

Pascale, R.M., Simile, M.M., De Miglio, M.R., Nufris, A., Seddaiu, M.A., Muroni, M.R. *et al.* (1995) Inhibition of 3-hydroxymethylglutaryl-CoA reductase activity and gene expression by dehydroepiandrosterone in preneoplastic liver nodules. *Carcinogenesis*, **16**, 1537–1542.

Peters, J.M., Zhou, Y.C., Ram, P.A., Lee, S.S.T., Gonzalez, F.J. and Waxman, D.J. (1996) Peroxisome proliferator-activated receptor alpha required for gene induction by dehydro-epiandrosterone-3 beta-sulfate. *Mol. Pharmacol.*, **50**, 67–74.

Prough, R.A., Webb, S.J., Wu, H.-Q., Lapensom, D.E. and Waxman, D.J. (1994) Induction of microsomal and peroxisomal enzymes by dehydroepiandrosterone and its reduced metabolite in rats. *Cancer Res.*, **54**, 2878–2886.

Ram, P.A. and Waxman, D.J. (1994) Dehydroepiandrosterone 3β-sulfate is an endogenous activator of the peroxisome-proliferation pathway: induction of cytochrome P-450 4A and acyl-CoA oxidase mRNAs in primary rat hepatocyte culture and inhibitory effects of Ca^{2+}-channel blockers. *Biochem. J.*, **301**, 753–758.

Rao, M.S., Musunuri, S. and Reddy, J.K. (1992a) Dehydroepiandrosterone-induced peroxisome proliferation in the rat liver. *Pathobiology*, **60**, 82–86.

Rao, M.S., Reid, B., Ide, H., Subbarao, V. and Reddy, J.K. (1994) Dehydroepiandrosterone-induced peroxisome proliferation in the rat: evaluation of sex differences. *Proc. Soc. Exp. Biol. Med.*, **207**, 186–190.

Rao, M.S. and Subbarao, V. (1998) Sex differences in dehydroepiandrosterone-induced hepatocarcinogenesis in the rat. *Cancer Lett.*, **125**, 111–116.

Rao, M.S., Subbarao, V., Kumar, S., Yeldandi, A.V. and Reddy, J.K. (1992b) Phenotypic properties of liver tumors induced by dehydroepiandrosterone in F-344 rats. *Jpn. J. Cancer Res.* **83**, 1179–1183.

Rao, M.S., Subbarao, V., Yeldandi, A.V. and Reddy, J.K. (1992c) Hepatocarcinogenicity of dehydroepiandrosterone in the rat. *Cancer Res.*, **52**, 2977–2979.

Reuter, S. and Mayer, D. (1995) Transport of dehydroepiandrosterone and dehydro-epiandrosterone sulphate into rat hepatocytes. *J. Steroid Biochem. Molec. Biol.*, **54**, 227–235.

Rose, K.A., Stapleton, G., Dott, K., Kieny, M.P., Best, R., Schwarz, M. *et al.* (1997). Cyp 7b, a novel brain cytochrome P450, catalyzes the synthesis of neurosteroids 7α-hydroxy-dehydro-epiandrosterone and 7α-hydroxy-pregnenolone. *Proc. Natl. Acad. Sci. USA*, **94**, 1925–1930.

Schriefers, H., Ghraf, R., Hoff, H.G. and Ockenfels, H. (1971) Einfluß von Alter und Geschlecht auf die Entwicklung und Differenzierung der Aktivitätsmuster von Enzymen des Steroidhormon-Stoffwechsels in der Leber von Ratten zweier verschiedener Tierstämme. *Hoppe-Seyler's Z. Physiol. Chem.*, **352**, 1363–1371.

Schwartz, A.G., Whitcomb, J.M., Nyce, J.W., Lewhart, M.L. and Pashko, L.L. (1988) Dehydro-epiandrosterone and structural analogs: a new class of cancer chemopreventive agents. *Adv. Cancer Res.*, **51**, 391–424.

Shepherd, A. and Cleary, M.P. (1984) Metabolic alterations after dehydroepiandrosterone treatment in Zucker rats. *Am. J. Physiol.*, **246**, 123–128.

Shibata, M.-A., Hasegawa, R., Imaida, K., Hagiwara, A., Ogawa, K., Hirose, M. *et al.* (1995) Chemoprevention by dehydroepiandrosterone and indomethacin in a rat multiorgan carcinogenesis model. *Cancer Res.*, **55**, 4870–4874.

Simile, A., Pascale, R.M., De Miglio, M.R., Nufris, A., Daino, L., Seddaiu, M.A. *et al.* (1995) Inhibition by dehydroepiandrosterone of growth and progression of persistent liver nodules in experimental rat liver carcinogenesis. *Int. J. Cancer*, **62**, 210–215.

Song, M.K., Grieco, D., Rall, J.E. and Nikodem, V.M. (1989) Thyroid hormone-mediated transcriptional activation of the rat liver malic enzyme gene by dehydroepiandrosterone. *J. Biol. Chem.*, **264**, 18981–18985.

Sourla, A., Martel, C., Labrie, C. and Labrie, F. (1998) Almost exclusive androgenic action of dehydroepiandrosterone in the rat mammary gland. *Endocrinology*, **139**, 753–764.

Starka, L. and Kutova, J. (1962) 7-Hydroxylation of dehydroepiandrosterone in rat liver homogenate. *Biochim. Biophys. Acta*, **56**, 76–82.

Svec, F. and Porter, J. (2000) Dehydroepiandrosterone and obesity. In M. Kalimi and W. Regelson (eds), *Dehydroepiandrosterone (DHEA), biochemical, physiological and clinical aspects*, Walter de Gruyter, Berlin, pp. 173–189.

Swierczynski, J., Bannasch, P. and Mayer, D. (1996) Increase of lipid peroxidation in rat liver microsomes by dehydroepiandrosterone feeding. *Biochim. Biophys. Acta*, **1315**, 193–198.

Swierczynski, J., Kochan, Z. and Mayer, D. (1997) Dietary α-tocopherol prevents DHEA-induced lipid peroxidation in rat liver microsomes and mitochondria. *Toxicol. Letters*, **91**, 129–136.

Swierczynski, J. and Mayer, D. (1996) Dehydroepiandrosterone-induced lipid peroxidation in rat liver mitochondria. *J. Steroid Biochem. Molec. Biol.*, **58**, 599–603.

Tagliaferro, A.R., Davis, J.R., Truchon, S. and van Hamont, N. (1986) Effects of dehydro-epiandrosterone acetate on metabolism, body weight and composition of male and female rats. *J. Nutr.*, **116**, 1977–1983.

Thornton, M., Moore, M.A. and Ito, N. (1989) Modifying influence of dehydroepiandro-sterone or butylated hydroxytoluene treatment on initiation and development stages of azaserine-induced acinar pancreatic lesions in the rat. *Carcinogenesis*, **10**, 407–410.

van Werden, E.M., Bierings, H.G., van Steenbrugge, G.J., de Jong, F.H. and Schröder, F.H. (1992) Adrenal glands of the mouse and rat do not synthesize androgens. *Life Sci.*, **50**, 857–861.

Waxman, D.J. (1996) Role of metabolism in the activation of dehydroepiandrosterone as a peroxisome proliferator. *J. Endocrinol.*, **150**, S129–S147.

Weber, E., Moore, M.A. and Bannasch, P. (1988a) Enzyme histochemical and morphological phenotype of amphophilic foci and amphophilic/tigroid cell adenomas in rat liver after combined treatment with dehydroepiandrosterone and *N*-nitrosomorpholine. *Carcinogenesis*, **9**, 1049–1054.

Weber, E., Moore, M.A. and Bannasch, P. (1988b) Phenotypic modulation of hepatocarcino-genesis and reduction in *N*-nitrosomorpholin-induced hemangiosarcoma and adrenal lesion development in Sprague-Dawley rats by dehydroepiandrosterone. *Carcinogenesis*, **9**, 1191–1195.

Wu, H., Masset-Brown, J., Tweedie, D.J., Milewich, L., Frenkel, R.A., Martin-Wixtrom, C. *et al*. (1989) Induction of microsomal NADPH-cytochrome P-450 reductase and cytochrome P-450 IVAI (P-450$_{LA}\omega$) by dehydroepiandrosterone in rats: a possible peroxisome proliferator. *Cancer Res.*, **49**, 2337–2343.

Yamada, J., Sakuma, M., Ikeda, T., Fukuda, K. and Suga, T. (1991) Characteristics of dehydroepiandrosterone as a peroxisome proliferator. *Biochim. Biophys. Acta*, **1092**, 233–243.

Yamada, J., Sakuma, M., Ikeda, T. and Suga, T. (1994) Activation of dehydroepiandrosterone as a peroxisome proliferator by sulfate conjugation. *Arch. Biochem. Biophys.*, **313**, 379–381.

Yen, T.T., Allan, J.A., Pearson, D.V. and Action, J.M. (1977) Prevention of obesity in Avy/a mice by dehydroisoandrosterone. *Lipids*, **12**, 409–413.

Zhang, H., Tamura, H., Yamada, J., Watanabe, T. and Suga, T. (1996) Effect of nicardipine calcium antagonist on induction of peroxisomal enzymes by dehydroepiandrosterone sulfate in cultured rat hepatocytes. *J. Toxicol. Sci.*, **21**, 235–241.

Chapter 6

DHEA and brain development

Synthia H. Mellon

DEFINITION OF A NEUROSTEROID

The intriguing observation that steroids could be synthesized in the brain came initially from experiments performed in the 1980's by Baulieu and colleagues who found that steroids such as pregnenolone, DHEA and their sulfate and lipoidal esters were present in higher concentrations in tissue from the nervous system (brain and peripheral nerve) than in the plasma. While these compounds could be due to peripheral synthesis and then sequestration in the brain, Baulieu and colleagues found that the steroids remained in the nervous system long after gonadectomy or adrenalectomy (Corpechot *et al.*, 1981, 1983), suggesting that steroids might either be synthesized *de novo* in the CNS and PNS or might accumulate in those structures. Such steroids were named "neurosteroids" to refer to their unusual origin and to differentiate them from steroids derived from more classical steroidogenic organs such as gonads, adrenals and placentae. How were these steroids synthesized? To test whether steroids were actually made in the brain or if they accumulated specifically in tissue from the nervous system, several laboratories, including our laboratory, determined directly if enzymes known to be involved in steroidogenesis adrenals, gonads and placentae could be responsible for neurosteroids synthesis. These results have established unequivocally that the enzymes found in classic steroidogenic tissues are indeed found in the nervous system (reviewed in Mensah-Nyagan *et al.*, 1999; Compagnone and Mellon, 2000).

But even if steroids are synthesized in the nervous system, (1) are they the same steroids that are typically secreted from the adrenals and gonads? (2) if not, what are they? (3) how do these neurosteroids function? Steroids have long been known to be specific ligands for intracellular (nuclear) steroid hormone receptors that bind to specific DNA sequences on hormone-responsive genes, and act as transcription factors to regulate the amount of transcription from the particular gene. At about the same time that Baulieu and colleagues had demonstrated that steroids could be synthesized *de novo* in the brain, neuropharmacologists demonstrated that certain steroids, specifically reduced derivatives of progesterone, could act as ligands for the ion-gated $GABA_A$ neurotransmitter receptor (Harrison and Simmonds, 1984; Majewska *et al.*, 1986). These progesterone derivatives caused an increase in the frequency and duration of $GABA_A$ channel opening, were effective at nanomolar concentrations, and appeared to modulate GABA activity at the $GABA_A$ receptor, rather than influencing receptor activity by themselves. Subsequent experiments by many laboratories demonstrated the steroid specificity of this binding and action (Harrison *et al.*, 1987; Wieland *et al.*, 1991). Following these studies, others demonstrated that other neurosteroids could also affect NMDA receptor activity, perhaps by a mechanism distinct from that for GABA$_A$ receptors (ffrench-Mullen and Spence, 1991; Wu *et al.*, 1991; Farb *et al.*, 1992; Park-Chung *et al.*, 1994). Finally, as another example of the mechanism by which neurosteroids affect gene regulation and cellular function, others have recently shown that steroids can affect neurotransmitter receptor function indirectly by affecting the membrane in which the receptor resides (Wetzel *et al.*, 1998). Thus, depending upon the steroid synthesized, neurosteroids could affect gene expression through action at classic intracellular nuclear receptors, or could affect neurotransmission through action at membrane ion-gated and other neurotransmitter receptors.

ENZYMES INVOLVED IN THE SYNTHESIS OF DHEA

Most of the enzymes present in the adrenal, gonad, and placenta have been found in the brain by measuring their enzymatic activity and/or their mRNA transcript level and/or their protein expression. The biosynthetic pathway of DHEA synthesis involves only two enzymes, shown in Figure 6.1. The first enzyme is P450scc, cholesterol side chain cleavage, a single enzyme that mediates the conversion of cholesterol to pregnenolone. The second reaction is mediated by P450c17, an enzyme that has both 17α hydroxylase and c17,20 lyase activities, and hence converts pregnenolone to 17α OH pregnenolone, and then to DHEA.

In the nervous system, there is not only region-specific expression of the steroidogenic enzymes, but there is also cell-type-specific and developmental regulation of these enzymes, indicating a more complex scheme than that depicted in the figure. The brain contains additional steroid metabolizing enzymes including sulfotransferases (HST) and sulfohydrolases (sulfatases, STS), that can interconvert DHEA and DHEAS, neuroactive compounds with different and distinct actions.

P450scc

The first, rate limiting, and hormonally regulated step in the synthesis of all steroid hormones is the conversion of cholesterol to pregnenolone. This reaction is catalyzed by the mitochondrial enzyme cholesterol side chain cleavage, P450scc (EC 1.14.15.6), in three successive chemical reactions: 20α-hydroxylation, 22-hydroxylation and scission of the c20–c22 carbon bond cholesterol. The products of this reaction are pregnenolone and isocaproic acid. A single P450scc species is found in all steroidogenic tissue, including the brain (Mellon and Deschepper, 1993; Mellon, 1994).

P450scc is the rate-limiting step in steroidogenesis, and is one of the slowest enzymes known, with a V_{max} of 1 mol cholesterol/mol enzyme/sec. The slowest part of this reaction may be the entry of cholesterol into the mitochondria and its binding to the active site of P450scc (*see the section on StAR*).

The human and rat genome contains a single gene encoding P450scc (Chung *et al.*, 1986; Matteson *et al.*, 1986a; Morohashi *et al.*, 1987; Oonk *et al.*, 1990), which is about 20 kb long, and contains nine exons. This gene encodes a mRNA of about

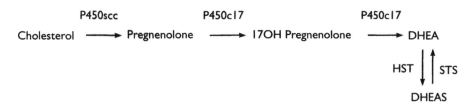

Figure 6.1 Schematic representation of the pathways leading to the synthesis of the neurosteroids DHEA and DHEAS from the precursor cholesterol. P450scc, cholesterol side chain cleavage; P450c17, 17α hydroxylase/c17,20 lyase, HST, sulfotransferase; STS, sulfatase.

2.0 kb, which encodes a 521 amino acid protein. This protein is proteolytically cleaved, removing a 39 amino acid leader peptide that directs the protein to the mitochondria. This protein is only active in the mitochondria (Black *et al.*, 1994). When P450scc, along with the electron donors adrenodoxin and adrenodoxin reductase, are directed to the microsomal fraction by use of a different leader peptide, P450scc is inactive, indicating that the inner mitochondrial milieu is absolutely required for activity.

Adrenodoxin reductase/adrenodoxin

P450scc functions as the terminal oxidase in a mitochondrial electron transport system. Electrons from NADPH are first accepted by a flavoprotein, adrenodoxin reductase, which is located in the mitochondrial matrix and is loosely associated with the inner membrane (Nakamura *et al.*, 1966; Omura *et al.*, 1966; Kimura and Suzuki, 1967). Adrenodoxin reductase transfers the electrons to an iron/sulfur protein, adrenodoxin, located in the mitochondrial matrix. Adrenodoxin first forms a complex with adrenodoxin reductase, dissociates after oxidation, and then binds to P450scc (or to the other mitochondrial P450s, P450c11β, P450c11AS and P450c11B3). These proteins are also often called "ferredoxin oxidoreductase" and "ferredoxin". In humans, there is one gene encoding adrenodoxin reductase, found on chromosome 17, and multiple functional adrenodoxin genes on chromosome 11, encoding identical mRNAs and proteins, and two nonfunctional adrenodoxin pseudogenes on chromosome 20. Adrenodoxin, but not adrenodoxin reductase, is transcriptionally regulated by tropic hormones, acting though cAMP.

StAR

A novel protein, steroidogenic acute regulatory protein, StAR, was identified as important for movement of cholesterol into the mitochondria in response to trophic hormone stimulation (Clark *et al.*, 1994; Stocco and Clark, 1996). In the adrenal and gonads, StAR expression and function are critical for steroidogenesis. StAR is a 30 kDa phosphoprotein that is synthesized as a 37 kDa precursor, containing an *N*-terminal mitochondrial targeting sequence. Individuals who are homozygous for mutations that inactivate StAR have a marked impairment in adrenal and gonadal steroidogenesis, but not in placental steroidogenesis (Lin *et al.*,

1993; Bose *et al.*, 1996). Patients with congenital lipoid adrenal hyperplasia do not appear to have gross neurological defects, suggesting that StAR may not be necessary for neurosteroidogenesis, as it is unnecessary for placental steroidogenesis (Saenger *et al.*, 1995). Nevertheless, StAR has been recently identified in the rodent brain, suggesting it may indeed play a role in regulating neurosteroidogenesis (Furukawa *et al.*, 1998).

P450c17

Pregnenolone and progesterone may undergo 17α hydroxylation to 17α-hydroxypregnenolone and 17α-hydroxyprogesterone. These steroids may then undergo scission of the c17,20 bond to form DHEA and androstenedione. These four reactions are all mediated by a single P450c17 enzyme (EC 1.14.99.9). P450c17 is bound to the smooth endoplasmic reticulum, and accepts electrons from P450 reductase. Since P450c17 has both 17α hydroxylase and 17,20 lyase activities, it catalyzes a key branch point in steroidogenesis. For example, in the human adrenal, the regional expression of P450c17 directs steroidogenesis to the production of mineralocorticoids, glucocorticoids or sex steroids. In the human zona glomerulosa, P450c17 is not expressed and pregnenolone is metabolized to mineralocorticoids; in the human zona fasciculata, P450c17 is expressed but the majority of the activity is 17α hydroxylase activity, and hence pregnenolone is directed to glucocorticoids; and in the zona reticularis, P450c17 has both 17α hydroxylase and c17,20 lyase activities, and hence pregnenolone is metabolized into sex steroids.

Several factors are important in determining whether a steroid will undergo 17,20 bond scission after 17 hydroxylation such as the presence of the cofactors (e.g. b5), and the potential competition for substrate between P450c17 and 3β HSD. In the human adrenal, 3βHSD mRNA and activity are low in the zona reticularis and high in the zona fasciculata (Voutilainen *et al.*, 1991; Coulter *et al.*, 1996; Endoh *et al.*, 1996). Cytochrome b5 also appears to have a great effect on c17,20 lyase activity (Auchus *et al.*, 1998). In the human adrenal, there appears to be a gradient of b5 expression in the human adrenal, with the highest concentration in the zona reticularis, indicating that P450c17 expressed in this zone would have greater lyase activity (see b5 section below) (Yanase *et al.*, 1998). Whether b5 expression is limited to particular regions of the nervous system, or in particular parts of neurons where P450c17 is expressed (i.e. in the soma versus in the neurites) is unknown, and may greatly influence the activity of this enzyme, and the subsequent steroid produced.

The 17α hydroxylase reaction occurs more readily than the 17,20 lyase reaction. P450c17 prefers Δ5 substrates, especially for 17,20 bond scission, accounting for the large concentrations of DHEA in the human adrenal. The single gene encoding P450c17 contains eight exons (Matteson *et al.*, 1986b).

P450 reductase

Both P450c17 and P450c21 receive electrons from a mitochondrial membrane-bound flavoprotein, P450 reductase, which is distinct from adrenodoxin reductase.

P450 reductase receives electrons from NADPH and transfers them one at a time to the microsomal P450. The second electron can also be provided by cytochrome b5 (see below). The amount of microsomal P450 reductase is less than that of both P450c21 and P450c17, resulting in competition for this protein. Therefore, factors that influence the association of a specific P450 with the reductase will likely influence the pathway that will be followed, e.g. c17 vs. c21 hydroxylation of progesterone or c17,20 lyase vs. c21 hydroxylation. Electron abundance also influences the activity of P450c17; increasing electron abundance favors both 17α hydroxylase *and* 17,20 lyase activities, while limiting electron abundance favors only 17α hydroxylase activity. Therefore, the ratio of P450 reductase to P450c17 seems to be critical in determining the pathway of steroidogenesis (Yanagibashi and Hall, 1986; Yamano *etal.*, 1989; Lin *etal.*, 1993).

Cytochrome b5

Cytochrome b5 is a small heme-containing protein which supplies electrons for many cytochrome P450-catalyzed reactions in the liver, and for the reduction of methemoglobin in erythrocytes. Cytochrome b5 is found both in a soluble form, necessary for its function in blood cells, and as a microsomal electron donor, playing the role of co-factor for microsomal P450s. Cytochrome b5 has been shown *in vitro*, to control the 17,20 lyase activity of P450c17 (Onoda and Hall, 1982; Kominami *etal.*, 1992; Auchus *etal.*, 1998). Cytochrome b5 specifically augments the 17,20 lyase activity of P450c17 *in vitro* and that this augmentation is produced by an allosteric modulation of P450c17, not through electron donor (Auchus *etal.*, 1998). Those results suggest that expression of b5 may be a mechanism by which c17,20 lyase activity, and hence DHEA production, is regulated in specific regions of the brain. There is precedent for this theory. In the human adrenal, b5 expression appears to be greater in the zona reticularis than in the zona fasciculata (Yanase *etal.*, 1998), and this differential expression may account for zone-specific synthesis of gluco-corticoids in the zona fasciculata vs. c19 steroids in the zona reticularis.

HST: sulfotransferase

Sulfation of free 3β hydroxy-steroids is a major enzymatic reaction of metabolism, excretion, and homeostasis of steroids and bile acids [reviewed in (Hobkirk 1985)]. Sulfation and sulfohydrolation activities have been reported in lung, kidney, adrenal, and testis, and have been described as crucial during development. 16-hydroxy DHEA-sulfate originating from the liver, DHEA-sulfate originating from the adrenal and DHEA originating from the placenta, serve as precursors for the production of estrone, estradiol and estriol in the developing human placenta (Iwamori *etal.*, 1976; Barker *etal.*, 1994; Parker *etal.*, 1994). Sulfotransferases are a family of cytosolic enzymes that conjugate steroid and phenolic substrates with inorganic sulfate derived from an active donor, PAPS (De Meio, 1975; Mulder, 1981). The hydroxysteroid sulfotransferase specifically uses Δ5 steroid substrates that are hydroxylated at C3, 5, 17, or 21 (De Meio, 1975). The resulting steroid sulfate esters are hydrophilic and are therefore more easily secreted from the cell. However, sulfation has a role greater than facilitating secretion, since it can also change the pharmacological activity of steroids, such as changing the way in which pregnenolone binds to the $GABA_A$ receptor [reviewed in (Majewska, 1991, 1992)]. HST is mainly found in adrenal in humans and in liver in other mammals (Ogura *et al.*, 1989).

STS: sulfatase

The steroid sulfohydrolase is a sterol-sulfate sulfohydrolase also known as steroid sulfatase or steroid-3-sulfatase. It specifically hydrolyzes sulfate groups in the 3β position of Δ5, steroids, such as pregnenolone, DHEA and androstenediol. STS is an important enzyme in steroid metabolism since its activity increases the pool of precursors that can be metabolized by other steroidogenic enzymes to produce biologically active sex steroids (De Meio, 1975; Shapiro, 1995). The human STS gene has been cloned and mapped to chromosome Xp22.3, proximal to the pseudoautosomal region (PAR), and the genetic aspect of this enzyme has been widely documented (Ballabio and Shapiro, 1995). Deficiency of STS activity results in severe ichthyosis caused by accumulation of steroid sulfates in the stratum corneum of the skin (Shapiro, 1982; Mohandas *et al.*, 1987). The rat and mouse STS have recently been cloned, and are fairly dissimilar. The mouse STS cDNA is 75% identical to the rat STS cDNA and only 63% identical to the human STS cDNA (Li *et al.*, 1996; Salido *et al.*, 1996).

DISTRIBUTION OF P450scc AND P450c17 IN THE BRAIN AND REGULATION OF THEIR EXPRESSION DURING DEVELOPMENT

The presence of functional steroidogenic enzymes in the brain has been established using enzymatic activity measurements, mRNA expression (using RT-PCR, ribonuclease protection assays, or *in situ* hybridization), and protein expression (using Western blotting or immunocytochemistry). These studies, by numerous laboratories over the years, were originally done mainly in the adult rat brain, but now include analysis of neurosteroidogenesis in the adult frog, bird and guinea-pig. There are only a limited number of reports that document steroidogenic enzyme expression during embryogenesis in rodents (Lauber and Lichtensteiger, 1994; Compagnone *et al.*, 1995a,b, 1997; Lauber and Lichtensteiger, 1996). However, such studies are crucial, since neurosteroids have diverse functions during development as well as in the adult.

Expression of P450scc in the adult rat brain

Initial support for the hypothesis that the brain is a steroidogenic organ derives from experiments demonstrating conversion of radioactive cholesterol to pregnenolone [reviewed in (Baulieu and Robel, 1990)]. While these experiments were suggestive that the brain had steroidogenic capacity, the demonstration that CNS tissue and neuronal cells contained P450scc mRNA and protein was hampered by the extremely low amount of this mRNA. Using RT/PCR and Southern blotting, we showed that P450scc mRNA was regionally expressed in extremely low abundance (Mellon and Deschepper, 1993). In the adult rat, it was found most abundantly in the cortex, and in a lesser extent in the amygdala, hippocampus and midbrain of both male and female rats. Purification of mixed primary glial cultures showed that Type-1 astrocytes synthesized P450scc mRNA (Mellon and Deschepper, 1993). Western blotting and immunocytochemistry showed that P450scc protein was almost as abundant in neonatal cultures of forebrain astrocytes as in mouse adrenocortical Y-1 cells, while P450scc mRNA was orders of magnitude less abundant suggesting that the protein was stable in the brain (Compagnone et al., 1995a).

Developmental regulation of P450scc expression in the CNS and PNS

P450scc mRNA and protein are expressed very early in development. By RNase protection assays, we detected P450scc mRNA as early as embryonic day 7.5 (E7.5) and found that its expression increased until E9.5. This increased expression was due to increased expression in the placenta (Durkee et al., 1992; Compagnone et al., 1995a). While P450scc mRNA could be readily detected in the developing gonads and adrenals, it could not be easily detected in the developing nervous system. However, by analyzing P450scc protein by immunocytochemistry, we demonstrated P450scc expression as early as E9.5 in the mouse (10.5 in the rat) in cells in the neural crest (Compagnone et al., 1995a). Expression of P450scc continued mainly in structures derived from the neural crest during embryogenesis and was found in the neuroepithelium, the retina, the trigeminal and dorsal root ganglia, in the neuroectoderm and in the thymus (Compagnone et al., 1995a). Consistent with a neural crest origin, cells expressing P450scc belong to several different cell lineages. P450scc was found in neurons of the dorsal root ganglia, trigeminal ganglia and in the neuroectoderm, and was found in glial lineages in the central nervous system (CNS). P450scc was also found in cells not derived from the neural crest, such as motor neurons (Compagnone et al., 1995a), Purkinje cells in the cerebellum (Ukena et al., 1998), in oligodendrocytes (Hu et al., 1987) and astrocytes (Mellon and Deschepper, 1993) in various regions of the brain from late embryogenesis to adulthood. In the peripheral nervous system (PNS), P450scc was expressed from E10.5 in condensing dorsal root ganglia and in cranial ganglia (Compagnone et al., 1995a). Other groups have also reported activity and expression of P450scc protein in peripheral nerves (Morfin et al., 1992). Thus, P450scc is expressed in a variety of cell types both in the CNS and the PNS. Its expression is initiated in the developing neural tube and in the neural crest, before organogenesis of the adrenal or of the gonads, suggesting a role for neurosteroidogenesis in neural development. Its expression in restricted areas of the brain do not seem to be developmentally regulated, as P450scc expression persists during adulthood.

Developmental regulation of P450c17 expression in the nervous system

Initial studies failed to detect expression of P450c17 mRNA in any region of the adult rat brain, using highly sensitive RT-PCR analyses (Mellon and Deschepper,

1993). Those studies also indicated that the steroidogenic enzyme proteins may be more easily detected than their mRNAs, since using an antibody directed against human P450c17, we determined the pattern of expression of P450c17 in developing rodent embryos (Compagnone *et al.*, 1995b). P450c17 expression is restricted in neurons in specific regions of the developing brain, and can be seen in both neuronal cell bodies and in fibers emanating from those cell bodies. We first observed P450c17 at E10.5 in cells migrating from the neural crest and condensing in dorsal root ganglia. Immunopositive cells were also observed in the neural tube in the position of the lateral motor column. Neuronal cell bodies immunopositive for P450c17 were restricted to the hindbrain mesopontine system (from E14.5), the thalamus and the neocortical subplate (from E16.5). Fibers extended in the areas of projections of these nuclei. We also found fibers coming from the trigeminal ganglion and the retina.

   In all the CNS regions studied, we found that P450c17 expression was transient and P450c17-immunostaining gradually disappeared in the first week of life. In the neonatal mouse and rat (day 0–7), P450c17 was mainly detected in the brainstem and in the cerebellum, internal capsule, olfactory tract, hippocampus, stria terminalis and thalamus. Few immunopositive cell bodies could be detected in any region of the adult rat brain, although fiber tracts could still readily be observed. In the PNS, P450c17 expression persisted in the adult, both in cell bodies and fibers. Thus, unlike P450scc, P450c17 is expressed in the CNS only during development, but persists in the PNS throughout life.

   DHEA was one of the first neurosteroids identified in the adult rodent brain, although at low concentrations. However, P450c17 mRNA, protein, or enzyme activity have not been found in cells of the CNS in adult animals. We propose that in the rodent, peripheral expression of P450c17 provides the brain with DHEA found in adult rat brains. An alternative pathway has been proposed for DHEA generation in the brain involving a Fe^{2+}-sensitive chemical reaction independent of P450 (Cascio *et al.*, 1998). However, it is unlikely that such a process occurs *in vivo* since the concentrations of Fe^{2+} necessary to induce production of DHEA (in the 10 mM range) in cells lacking P450c17 is far out of a physiological range. We thus believe that, in the adult, DHEA is delivered to the spinal cord and brain via the peripheral nerves and terminals which express P450c17 and synthesize DHEA.

DEVELOPMENTAL ROLE OF DHEA AND DHEAS IN ORGANOGENESIS

DHEA and DHEAS are modulators of cytoarchitectural organization in the developing brain

What function do DHEA and DHEAS play during development, when they appear to be synthesized by discrete regions of the nervous system? We hypothesized that DHEA, synthesized in neurons that express P450c17, may modulate those neurons directly, or may modulate neurons to which P450c17-expressing neurons form synapses. One particular region in which P450c17 was expressed in the embryo was the neocortical subplate. This region of the developing CNS has been reported to be the region that receives thalamic projections, produces signals for cortical projections, and may produce signals for efferent thalamic projections from the cortex (Miyashita-Lin *et al.*, 1999; Rubenstein *et al.*, 1999). Hence, we hypothesized that DHEA synthesized in that region, may be a signal that induces axonal guidance of cortical projection. To determine if DHEA or DHEAS had such effects, or more simply, could cause specific neurite growth, we cultured embryonic neocortical neurons to determine the effects of DHEA and DHEAS on these cultures. Previous studies had shown that DHEA and DHEAS could induce neurite growth in mixed neuronal/glial cultures (Bologa *et al.*, 1987), but whether DHEA and DHEAS acted on neurons directly, or on glia was unknown. Furthermore, the nature of the neurite growth was unknown. Our experiments, using pure neuronal cultures in serum-supplement medium, demonstrated that DHEA selectively increased Tau-1 immunopositive (axonal) outgrowth, while DHEAS selectively increased MAP-2 immunopositive (dendritic) growth (Compagnone and Mellon, 1998). We further determined that, in the developing neocortex, DHEA promoted other morphological indices of synaptic contacts while DHEAS promoted branching, suggesting that these two neurosteroids are separate hormones active in the same process. Thus, the local ratio of DHEA versus DHEAS may regulate specific neurite growth in the developing neocortex, thereby shaping projections and synapses during embryogenesis.

We also hypothesized that the sulfohydrolase (STS) and sulfotransferase (HST) activities in the developing brain are key regulators of DHEAS effects in the developing brain. We identified STS mRNA expression in the thalamus, site where P450c17 fibers terminate and where P450c17 is active during embryogenesis (Compagnone *et al.*, 1997). The thalamus was identified as a region involved in pioneering axons to the neocortex (Shatz *et al.*, 1990). The presence of the sulfohydrolase in this region is consistent with the presence of axonal growth promoting cues in the thalamus. Others have shown that allopregnanolone induced axonal regression in the developing hippocampus (Brinton, 1994) suggesting that neurosteroids produced *de novo* in the nervous system may work in concert to promote, guide and refine axonal growth and synaptic connections in the developing neocortex. The small number of responsive cells within the cultures suggested that DHEA was active on only a few neurons in our culture, and these may be subplate neurons that express P450c17.

To identify downstream events induced by DHEA/DHEAS in primary neocortical cells, and to determine if these neurosteroids could ultimately alter the expression of certain mRNAs and proteins involved in neuronal function, we performed semi-quantitative RT-PCR in control cells and in cells treated with DHEA (10^{-12} M) or DHEAS (10^{-10} M) for 16 hours (Compagnone and Mellon, 2000). Since we had demonstrated that DHEA increased Tau-1 expression, a microtubule associated protein segregated in axons, it was not unexpected that we found that DHEA induced expression of Tau mRNA. Similarly, since we had shown that DHEAS increased MAP-2 protein expression, a microtubule associated protein segregated in dendrites, it was also not surprising that we found that DHEA-S induced MAP-2 mRNA. Furthermore, we also found that DHEA induced the dopamine receptor Type 1 and 2 mRNAs while DHEAS reduced the expression of both these mRNAs to undetectable levels, again demonstrating that DHEA and DHEAS act as distinct neurosteroids. These data suggest that both DHEA and DHEAS enhanced the differentiation of the cultured neocortical neurons, but each steroid produced specific downstream events. Taken together, our results suggest that DHEA is strongly involved in axonal growth and functional activation of specific neuronal networks and that at least part of these effects are mediated via NMDA receptor activation. We believe that DHEA may serve a similar role in all the regions where P450c17 is expressed and active.

DHEA modulates NMDA receptors

Neuroactive steroids such as pregnenolone-sulfate, DHEA and DHEA-sulfate, have been shown to modulate NMDA receptor activity (Wu et al., 1991; Bowlby, 1993; Park-Chung et al., 1994; Fahey et al., 1995; Compagnone and Mellon, 1998). Allopregnanolone-sulfate, but not allopregnanolone, acts as a negative allosteric modulator of NMDA receptors (Park-Chung et al., 1994) while DHEA, pregnenolone, and their sulfate esters are considered positive allosteric modulators of NMDA receptors. Unlike GABA$_A$ receptor interactions, the interaction of neurosteroids with the NMDA receptor is not well documented, and no specific interaction sites have been described. Molecular cloning has demonstrated the presence of five NMDA receptor subunits, called $\zeta 1$ in the mouse (NR1 in the rat), and four ε subunits (1–4) in the mouse (NR2A-D in the rat) (Moriyoshi et al., 1991; Ikeda et al., 1992; Kutsuwada et al., 1992; Meguro et al., 1992; Monyer et al., 1992; Yamazaki et al., 1992). The $\zeta 1$ subunit is obligatory for channel function and different splice variants in its mRNA have been found (Moriyoshi et al., 1991). The ε subunit mRNAs are expressed in a varying regional and developmental distribution, suggesting that differences in the NMDA receptor composition in various regions of the brain and during early postnatal development in the cerebellum may result in different pharmacologic profiles.

In a pure neuronal primary culture of mouse embryonic neocortical neurons, we showed that DHEA, but not DHEAS, could evoke an increase in free intracellular calcium in the absence of KCl or NMDA (Compagnone and Mellon, 1998). DHEAS potentiated glutamate and NMDA-mediated increases in free intracellular calcium concentrations in similar primary neuronal chick cultures (Fahey et al., 1995). DHEA-mediated increases in free intracellular calcium concentrations occurred at low concentrations of DHEA (from 10^{-12}M) in high density cell suspensions (Compagnone and Mellon, 1998) and from 10^{-8}M in low density primary neuronal cultures. We also showed that DHEA, as others had shown for pregnenolone sulfate, greatly potentiated NMDA induced-increases in free intracellular calcium concentrations. Increased intracellular calcium mediated by DHEA was abolished in a dose dependent manner by both MK801, a non-competitive antagonist of NMDA, and by D-AP5, a competitive inhibitor of NMDA receptor, suggesting a direct interaction of DHEA at the NMDA receptor. These data demonstrated for the first time, the ability of DHEA to act as a neuromodulator at NMDA receptors. In our experiments, embryonic cortical neurons were also mildly responsive to glycine since 10 μM glycine increased the free intracellular calcium concentration by 2.6-fold. DHEA had an additive and not a potentiating effect on glycine-mediated increases in free intracellular calcium concentrations, suggesting that DHEA and glycine acted independently at the NMDA receptor.

Which subunits mediate the DHEA effect, and are there specific NMDA subunit combinations that specifically mediate the effect of neurosteroids? To begin answering these questions, we performed experiments to determine which NMDA subunits were present in our cultures. The rather low affinity of the NMDA receptors for glycine and their sensitivity for D-AP5 suggested the presence of the subunit $\epsilon1$ in our embryonic neocortical cultures (Anson *et al.*, 1998; Kew *et al.*, 1998). Using RT-PCR, we confirmed that both $\epsilon1$ and $\epsilon2$ mRNAs were expressed in our cultures. Others had shown that NR2A expression was developmentally regulated in neuronal cortical cultures from different ages of neonatal rats (Kew *et al.*, 1998). The late expression of NR2A was correlated with the appearance of a markedly reduced affinity for glycine around the 20th postnatal day.

In the cerebellum, the expression of NR2 subtype mRNA correlated with the developmental migration and postmigratory maturation of granule cells. During migration and immediately after arrival in the internal granular cell layer, NR2B was the only subunit expressed, and its expression declined to undetectable levels in the adult (Monyer *et al.*, 1994). Expression of NR2A mRNA correlated with the formation of synaptic contacts in the cerebellum and was heavily expressed during synaptic pruning. More recently, it was shown that receptors containing both $\epsilon1$ and $\epsilon3$ subunits could be present at the synapse from granule cells in cerebellar slices (Takahashi *et al.*, 1996). Our results obtained from semi-quantitative RT-PCR suggested that the $\epsilon2$ subunit was down regulated by increased concentrations of DHEA. Thus DHEA may be involved in inducing plasticity in NMDA receptor subunit composition.

It is known that certain molecules that bind directly to NMDA receptors or potentiate its effects, regulate its subunit composition. Examples include chronic ethanol treatment (Kumari and Ticku, 1998), which selectively increased NR2B mRNA in the cortex without altering NR1 mRNA. Similarly, treatment with the NMDA antagonist D-AP5 produced changes in NMDA receptor subunit composition in cultured mouse cortical neurons. D-AP5 selectively increased NR2B mRNA, modifying the ratio of NR2B/2A proteins. It also increased NR1 protein without modifying NR1 mRNA levels (Follesa and Ticku, 1996). Hence it is likely that other molecules that bind directly to NMDA receptors or potentiate its effects, like DHEA, may also regulate its subunit composition in a particular location, resulting in a further modulation of these receptors.

Behavioral effects of neurosteroids such as pregnenolone sulfate and DHEAS in enhancing spatial memory in mice are thought to result from their positive modulatory action at the NMDA receptor (Mathis *et al.*, 1994). This effect may also result from changes in NMDA receptor subunit composition. Both $\epsilon1$ and $\epsilon2$ NMDA receptor subunits are expressed in hippocampal CA1 (Sakimura *et al.*, 1995) and CA3 neurons (Ito *et al.*, 1997). In $\epsilon1$ knock-out mice, impairment of $\epsilon1$ expression caused loss in long term potentiation at the hippocampal CA1 synapses associated with deficiency in spatial learning (Sakimura *et al.*, 1995) and resulted in a reduction of NMDA EPSCs and LTP in response to the commissural/associational-CA3 but not fimbrial-CA3 synapses (Ito *et al.*, 1997). On the other hand, $\epsilon2$ impairment produced reversed effects in CA3 neurons, reducing NMDA EPSCs and LTP in fimbrial-CA3 synapses with no appreciable modification of the commissural/associ-ational-CA3 synapses. These results suggest that in addition to changes in NMDA subunit composition, function of NMDA receptor is regulated in a synapse selective manner. Hence, DHEA may also affect NMDA responsiveness and subunit compos-ition in a synapse-selective manner as well. Mapping of the regions of the developing CNS and PNS that are responsive to DHEA and DHEAS will greatly aid our ability to understand the diverse functions of these two distinct neuromodulators.

ACKNOWLEDGMENTS

This work was supported by grants from the NIH, March of Dimes, and Alzheimer's Association.

REFERENCES

Anson, L.C., Chen, P.E., Wyllie, D.J.A., Colquhoun, D. and Schoepfer, R. (1998) Identification of amino acid residues of the NR2A subunit that control glutamate potency in recombinant NR1/NR2A NMDA receptors. *J. Neurosci.*, **18**, 581–589.

Auchus, R.J., Lee, T.C. and Miller, W.L. (1998) Cytochrome b5 augments the 17,20-lyase activity of human P450c17 without direct electron transfer. *J. Biol. Chem.*, **273**, 3158–3165.

Ballabio, A. and Shapiro, L.J. (1995) Steroid sulfatase deficiency and X-linked ichtyosis. In C.R. Scriver, A.L. Beaudet, W.S. Sly and D. Valle (eds), *The metabolic and molecular bases of inherited disease*, New York: Mc Graw-Hill, pp. 2999–3022.

Barker, E.V., Hume, R., Hallas, A. and Coughtrie, W.H. (1994) Dehydroepiandrosterone sulfotransferase in the developing human fetus: quantitative biochemical and immunological characterization of the hepatic, renal, and adrenal enzyme. *Endocrinology*, **134**, 982–989.

Baulieu, E.E. and Robel, P. (1990) Neurosteroids: a new brain function? *J. Steroid Biochem. Mol. Biol.*, **37**, 395–403.

Black, S.M., Harikrishna, J.A., Szklarz, G.D. and Miller, W.L. (1994) The mitochondrial environment is required for activity of the cholesterol side-chain cleavage enzyme, cyto-chrome P450scc. *Proc. Natl. Acad. Sci. USA*, **91**, 7247–7251.

Bologa, L., Sharma, J. and Roberts, E. (1987) Dehydroepiandrosterone and its sulfated derivative reduce neuronal death and enhance astrocytic differentiation in brain cell cultures. *J. Neurosci. Res.*, **17**, 225–234.

Bose, H.S., Sugawara, T., Strauss, J.F.R. and Miller, W.L. (1996) The pathophysiology and genetics of congenital lipoid adrenal hyperplasia. International congenital lipoid adrenal hyperplasia consortium. *N. Engl. J. Med.*, **335**, 1870–1878.

Bowlby, M.R. (1993) Pregnenolone sulfate potentiation of N-methyl-D-aspartate receptor channels in hippocampal neurons. *Mol. Pharmacol.*, **43**, 813–819.

Brinton, R.D. (1994) The neurosteroid 3 alpha-hydroxy-5 alpha-pregnan-20-one induces cyto-architectural regression in cultured fetal hippocampal neurons. *J. Neurosci.*, **14**, 2763–2774.

Cascio, C., Prasad, V.V., Lin, Y.Y., Lieberman, S. and Papadopoulos, V. (1998) Detection of P450c17-independent pathways for dehydroepiandrosterone (DHEA) biosynthesis in brain glial tumor cells. *Proc. Natl. Acad. Sci. USA*, **95**, 2862–2867.

Chung, B.C., Matteson, K.J., Voutilainen, R., Mohandas, T.K. and Miller, W.L. (1986) Human cholesterol side-chain cleavage enzyme, P450scc: cDNA cloning, assignment of the gene to chromosome 15, and expression in the placenta. *Proc. Natl. Acad. Sci. USA*, **83**, 8962–8966.

Clark, B.J., Wells, J., King, S.R. and Stocco, D.M. (1994) The purification, cloning, and expression of a novel luteinizing hormone-induced mitochondrial protein in MA-10 mouse Leydig tumor cells. Characterization of the steroidogenic acute regulatory protein (StAR). *J. Biol. Chem.*, **269**, 28314–28322.

Compagnone, N.A., Bulfone, A., Rubenstein, J.L. and Mellon, S.H. (1995a) Expression of the steroidogenic enzyme P450scc in the central and peripheral nervous systems during rodent embryogenesis. *Endocrinology*, **136**, 2689–2696.

Compagnone, N.A., Bulfone, A., Rubenstein, J.L. and Mellon, S.H. (1995b) Steroidogenic enzyme P450c17 is expressed in the embryonic central nervous system. *Endocrinology*, **136**, 5212–5223.

Compagnone, N.A. and Mellon, S.H. (1998) Dehydroepiandrosterone: a potential signalling molecule for neocortical organization during development. *Proc. Natl. Acad. Sci. USA*, **95**, 4678–4683.

Compagnone, N.A. and Mellon, S.H. (2000) Neurosteroids: biosynthesis and function of these novel neuromodulators. *Front. Neuroendocrin.*, **21**, 1–58.

Compagnone, N.A., Salido, E., Shapiro, L.J. and Mellon, S.H. (1997) Expression of steroid sulfatase during embryogenesis. *Endocrinology*, **138**, 4768–4773.

Corpechot, C., Robel, P., Axelson, M., Sjovall, J. and Baulieu, E.E. (1981) Characterization and measurement of dehydroepiandrosterone sulfate in rat brain. *Proc. Natl. Acad. Sci. USA*, **78**, 4704–4707.

Corpechot, C., Synguelakis, M., Talha, S., Axelson, M., Sjovall, J., Vihko, R. *et al.* (1983) Pregnenolone and its sulfate ester in the rat brain. *Brain Res.*, **270**, 119–125.

Coulter, C.L., Goldsmith, P.C., Mesiano, S., Voytek, C.C., Martin, M.C. *et al.* (1996) Functional maturation of the primate fetal adrenal *in vivo*. II. Ontogeny of corticosteroid synthesis is dependent upon specific zonal expression of 3 beta-hydroxysteroid dehydrogenase/isomerase. *Endocrinology*, **137**, 4953–4959.

De Meio, R. (1975) Sulfate activation and transfert. *Metabolism of sulfur compounds*, New York: Academic Press, pp. 287–359.

Durkee, T.J., McLean, M.P., Hales, D.B., Payne, A.H., Waterman, M.R., Khan, I. *et al.* (1992) P450(17 alpha) and P450scc gene expression and regulation in the rat placenta. *Endocrinology*, **130**, 1309–1317.

Endoh, A., Kristiansen, S.B., Casson, P.R., Buster, J.E. and Hornsby, P.J. (1996) The zona reticularis is the site of biosynthesis of dehydroepiandrosterone and dehydroepiandrosterone sulfate in the adult human adrenal cortex resulting from its low expression of 3 beta-hydroxysteroid dehydrogenase. *J. Clin. Endocrinol. Metab.*, **81**, 3558–3565.

Fahey, J.M., Lindquist, D.G., Pritchard, G.A. and Miller, L.G. (1995) Pregnenolone sulfate potentiation of NMDA-mediated increases in intracellular calcium in cultured chick cortical neurons. *Brain Res.*, **669**, 183–188.

Farb, D.H., Gibbs, T.T., Wu, F.S., Gyenes, M., Friedman, L. and Russek, S.J. (1992) Steroid modulation of amino acid neurotransmitter receptors. *Adv. Biochem. Psychopharmacol.*, **47**, 119–131.

ffrench-Mullen, J.M. and Spence, K.T. (1991) Neurosteroids block Ca^{2+} channel current in freshly isolated hippocampal CA1 neurons. *Eur. J. Pharmacol.*, **202**, 269–272.

Follesa, P. and Ticku, M.K. (1996) NMDA receptor upregulation: molecular studies in cultured mouse cortical neurons after chronic antagonist exposure. *J. Neurosci.*, **16**, 2172–2178.

Furukawa, A., Miyatake, A., Ohnishi, T. and Ichikawa, Y. (1998) Steroidogenic acute regulatory protein (StAR) transcripts constitutively expressed in the adult rat central nervous system: colocalization of StAR, cytochrome P-450SCC (CYP XIA1), and 3beta-hydroxysteroid dehydrogenase in the rat brain. *J. Neurochem.*, **71**, 2231–2238.

Harrison, N.L., Majewska, M.D., Harrington, J.W. and Barker, J.L. (1987) Structure–activity relationships for steroid interaction with the gamma-aminobutyric acidA receptor complex. *J. Pharmacol. Exp. Ther.*, **241**, 346–353.

Harrison, N.L. and Simmonds, M.A. (1984) Modulation of GABA receptor complex by a steroid anesthetic. *Brain Res.*, **323**, 284–293.

Hobkirk, R. (1985) Steroid sulfotransferases and steroid sulfate sulfatases: characteristics and biological roles. *Can. J. Biochem. Cell. Biol.*, **63**, 1127–1144.

Hu, Z.Y., Bourreau, E., Jung-Testas, I., Robel, P. and Baulieu, E.E. (1987) Neurosteroids: oligodendrocyte mitochondria convert cholesterol to pregnenolone. *Proc. Natl. Acad. Sci. USA*, **84**, 8215–8219.

Ikeda, K., Nagasawa, M., Mori, H., Araki, K., Sakimura, K., Watanabe, M. *et al.* (1992) Cloning and expression of the epsilon 4 subunit of the NMDA receptor channel. *FEBS Lett.*, **313**, 34–38.

Ito, I., Futai, K., Katagiri, H., Watanabe, M., Sakimura, K., Mishina, M. *et al.* (1997) Synapse-selective impairment of NMDA receptor functions in mice lacking NMDA receptor epsilon 1 or epsilon 2 subunit. *J. Physiol. (Lond.)*, **500**, 401–408.

Iwamori, M., Moser, H.W. and Kishimoto, Y. (1976) Steroid sulfatase in brain: comparison of sulfohydrolase activities for various steroid sulfates in normal and pathological brains, including the various forms of metachromatic leukodystrophy. *J. Neurochem.*, **27**, 1389–1395.

Kew, J.N., Richards, J.G., Mutel, V. and Kemp, J.A. (1998) Developmental changes in NMDA receptor glycine affinity and ifenprodil sensitivity reveal three distinct populations of NMDA receptors in individual rat cortical neurons. *J. Neurosci.*, **18**, 1935–1943.

Kimura, T. and Suzuki, K. (1967) Components of the electron transport system in adrenal steroid hydroxylase. *J. Biol. Chem.*, **242**, 485–491.

Kominami, S., Ogawa, N., Morimune, R., De-Ying, H. and Takemori, S. (1992) The role of cytochrome b5 in adrenal microsomal steroidogenesis. *J. Steroid Biochem. Mol. Biol.*, **42**, 57–64.

Kumari, M. and Ticku, M.K. (1998) Ethanol and regulation of the NMDA receptor subunits in fetal cortical neurons. *J. Neurochem.*, **70**, 1467–1473.

Kutsuwada, T., Kashiwabuchi, N., Mori, H., Sakimura, K., Kushiya, E., Araki, K. *et al.* (1992) Molecular diversity of the NMDA receptor channel [see comments]. *Nature*, **358**, 36–41.

Lauber, M.E. and Lichtensteiger, W. (1994) Pre- and postnatal ontogeny of aromatase cytochrome P450 messenger ribonucleic acid expression in the male rat brain studied by *in situ* hybridization. *Endocrinology*, **135**, 1661–1668.

Lauber, M.E. and Lichtensteiger, W. (1996) Ontogeny of 5 alpha-reductase (type 1) messenger ribonucleic acid expression in rat brain: early presence in germinal zones. *Endocrinology*, **137**, 2718–2730.

Li, X.M., Salido, E.C., Gong, Y., Kitada, K., Serikawa, T., Yen, P.H. *et al.* (1996) Cloning of the rat steroid sulfatase gene (Sts), a non-pseudoautosomal X-linked gene that undergoes X inactivation. *Mamm. Genome*, **7**, 420–424.

Lin, D., Black, S.M., Nagahama, Y. and Miller, W.L. (1993) Steroid 17 alpha-hydroxylase and 17,20-lyase activities of P450c17: contributions of serine106 and P450 reductase. *Endocrinology*, **132**, 2498–2506.

Majewska, M. (1991) Neurosteroids: GABAA-agonistic and GABAA-antagonistic modulators of the GABAA receptor. In E. Costa and S.M. Paul (eds), *Neurosteroids and brain function*, New York: Thieme, pp. 109–117.

Majewska, M.D. (1992) Neurosteroids: endogenous bimodal modulators of the GABAA receptor. Mechanism of action and physiological significance. *Prog. Neurobiol.*, **38**, 379–395.

Majewska, M.D., Harrison, N.L. and Schwartz, R.D. (1986) Steroid hormone metabolites are barbiturate-like modulators of the GABA receptor. *Science*, **232**, 1004–1007.

Mathis, C., Paul, S.M. and Crawley, J.N. (1994) The neurosteroid pregnenolone sulfate blocks NMDA antagonist-induced deficits in a passive avoidance memory task. *Psychopharmacology (Berl.)*, **116**, 201–206.

Matteson, K.J., Chung, B.C., Urdea, M.S. and Miller, W.L. (1986a) Study of cholesterol side-chain cleavage (20,22 desmolase) deficiency causing congenital lipoid adrenal hyperplasia using bovine-sequence P450scc oligodeoxyribonucleotide probes. *Endocrinology*, **118**, 1296–1305.

Matteson, K.J., Picado-Leonard, J., Chung, B.C., Mohandas, T.K. and Miller, W.L. (1986b) Assignment of the gene for adrenal P450c17 (steroid 17 alpha-hydroxylase/17,20 lyase) to human chromosome 10. *J. Clin. Endocrinol. Metab.*, **63**, 789–791.

Meguro, H., Mori, H., Araki, K., Kushiya, E., Kutsuwada, T., Yamazaki, M. *et al.* (1992) Functional characterization of a heteromeric NMDA receptor channel expressed from cloned cDNAs. *Nature*, **357**, 70–74.

Mellon, S.H. (1994) Neurosteroids: biochemistry, modes of action, and clinical relevance. *J. Clin. Endocrinol. Metab.*, **78**, 1003–1008.

Mellon, S.H. and Deschepper, C.F. (1993) Neurosteroid biosynthesis: genes for adrenal steroidogenic enzymes are expressed in the brain. *Brain Res.*, **629**, 283–292.

Mensah-Nyagan, A.G., Do-Rego, J.L., Beaujean, D., Luu-The, V., Pelletier, G. and Vaudry, H. (1999) Neurosteroids: expression of steroidogenic enzymes and regulation of steroid biosynthesis in the central nervous system. *Pharmacol. Rev.*, **51**, 63–81.

Miyashita-Lin, E.M., Hevner, R., Wassarman, K.M., Martinez, S. and Rubenstein, J.L. (1999) Early neocortical regionalization in the absence of thalamic innervation. *Science*, **285**, 906–909.

Mohandas, T., Geller, R.L., Yen, P.H., Rosendorff, J., Bernstein, R., Yoshida, A. *et al.* (1987) Cytogenetic and molecular studies on a recombinant human X chromosome: implications for the spreading of X chromosome inactivation. *Proc. Natl. Acad. Sci. USA*, **84**, 4954–4958.

Monyer, H., Burnashev, N., Laurie, D.J., Sakmann, B. and Seeburg, P.H. (1994) Developmental and regional expression in the rat brain and functional properties of four NMDA receptors. *Neuron*, **12**, 529–540.

Monyer, H., Sprengel, R., Schoepfer, R., Herb, A., Higuchi, M., Lomeli, H. *et al.* (1992) Heteromeric NMDA receptors: molecular and functional distinction of subtypes. *Science*, **256**, 1217–1221.

Morfin, R., Young, J., Corpechot, C., Egestad, B., Sjovall, J. and Baulieu, E.E. (1992) Neurosteroids: pregnenolone in human sciatic nerves. *Proc. Natl. Acad. Sci. USA*, **89**, 6790–6793.

Moriyoshi, K., Masu, M., Ishii, T., Shigemoto, R., Mizuno, N. and Nakanishi, S. (1991) Molecular cloning and characterization of the rat NMDA receptor (see comments). *Nature*, **354**, 31–37.

Morohashi, K., Sogawa, K., Omura, T. and Fujii-Kuriyama, Y. (1987) Gene structure of human cytochrome P-450(SCC), cholesterol desmolase. *J. Biochem. (Tokyo)*, **101**, 879–887.

Mulder, G. (1981) *The sulfatation of drugs and other compounds*. Boca Raton: CRC press.

Nakamura, Y., Otsuka, H. and Tamaoki, B. (1966) Requirement of a new flavoprotein and a non-heme iron-containing protein in the steroid 11β- and 18-hydroxylase system. *Biochem. Biophys. Acta*, **122**, 34–42.

Ogura, K., Kajita, J., Narihata, H., Watabe, T., Ozawa, S., Nagata, K. *et al.* (1989) Cloning and sequence analysis of a rat liver cDNA encoding hydroxysteroid sulfotransferase. *Biochem. Biophys. Res. Commun.*, **165**, 168–174.

Omura, T., Sanders, S., Estabrook, R.W., Cooper, D.Y. and Rosenthal, O. (1966) Isolation from adrenal cortex of a non-heme iron protein and a flavoprotein functional as a reduced triphosphopyridine nucleotide-cytochrome P-450 reductase. *Arch. Biochem. Biophys.*, **117**, 660.

Onoda, M. and Hall, P.F. (1982) Cytochrome b5 stimulates purified testicular microsomal cytochrome P-450 (C21 side-chain cleavage). *Biochem. Biophys. Res. Commun.*, **108**, 454–460.

Oonk, R.B., Parker, K.L., Gibson, J.L. and Richards, J.S. (1990) Rat cholesterol side-chain cleavage cytochrome P-450 (P-450scc) gene. Structure and regulation by cAMP *in vitro*. *J. Biol. Chem.*, **265**, 22392–22401.

Park-Chung, M., Wu, F.S. and Farb, D.H. (1994) 3 alpha-Hydroxy-5 beta-pregnan-20-one sulfate: a negative modulator of the NMDA-induced current in cultured neurons. *Mol. Pharmacol.*, **46**, 146–150.

Parker, Jr. C.R., Falany, C.N., Stockard, C.R., Stankovic, A.K. and Grizzle, W.E. (1994) Immunohistochemical localization of dehydroepiandrosterone sulfotransferase in human fetal tissues. *J. Clin. Endocrinol. Metab.*, **78**, 234–236.

Rubenstein, J.L., Anderson, S., Shi, L., Miyashita-Lin, E., Bulfone, A. and Hevner, R. (1999) Genetic control of cortical regionalization and connectivity. *Cereb. Cortex*, **9**, 524–532.

Saenger, P., Klonari, Z., Black, S.M., Compagnone, N., Mellon, S.H., Fleischer, A. *et al.* (1995) Prenatal diagnosis of congenital lipoid adrenal hyperplasia. *J. Clin. Endocrinol. Metab.*, **80**, 200–205.

Sakimura, K., Kutsuwada, T., Ito, I., Manabe, T., Takayama, C., Kushiya, E. *et al.* (1995) Reduced hippocampal LTP and spatial learning in mice lacking NMDA receptor epsilon 1 subunit. *Nature*, **373**, 151–155.

Salido, E.C., Li, X.M., Yen, P.H., Martin, N., Mohandas, T.K. and Shapiro, L.J. (1996) Cloning and expression of the mouse pseudoautosomal steroid sulphatase gene (Sts). *Nat. Genet.*, **13**, 83–86.

Shapiro, I. (1982) Steroid sulfatase deficiency and X-linked ichthyosis. In J.W. Stanbury, J.B. Wyngaarden, D.S. Fredrickson, J.L., Goldstein, M.S. Brown (eds). *The metabolic basis of inherited diseases*, New York: McGraw Hill, pp. 1027–1034.

Shapiro, L. (1995) Steroid sulfatase. In International Symposium on DHEA transformation into androgens and estrogens in target tissues: Intracrinology (Québec City).

Shatz, C.J., Ghosh, A., McConnell, S.K., Allendoerfer, K.L., Friauf, E. and Antonini, A. (1990) Pioneer neurons and target selection in cerebral cortical development. *Cold Spring Harb. Symp. Quant. Biol.*, **55**, 469–480.

Stocco, D.M. and Clark, B.J. (1996) Regulation of the acute production of steroids in steroidogenic cells. *Endocr. Rev.*, **17**, 221–244.

Takahashi, T., Feldmeyer, D., Suzuki, N., Onodera, K., Cull-Candy, S.G., Sakimura, K. *et al.* (1996) Functional correlation of NMDA receptor epsilon subunits expression with the properties of single-channel and synaptic currents in the developing cerebellum. *J. Neurosci.*, **16**, 4376–4382.

Ukena, K., Usui, M., Kohchi, C. and Tsutsui, K. (1998) Cytochrome P450 side-chain cleavage enzyme in the cerebellar Purkinje neuron and its neonatal change in rats. *Endocrinology*, **139**, 137–147.

Voutilainen, R., Ilvesmaki, V. and Miettinen, P.J. (1991) Low expression of 3 beta-hydroxy-5-ene steroid dehydrogenase gene in human fetal adrenals *in vivo*; adrenocorticotropin and protein kinase C-dependent regulation in adrenocortical cultures. *J. Clin. Endocrinol. Metab.*, **72**, 761–767.

Wetzel, C.H., Hermann, B., Behl, C., Pestel, E., Rammes, G., Zieglgansberger, W. *et al.* (1998) Functional antagonism of gonadal steroids at the 5-hydroxytryptamine type 3 receptor. *Mol. Endocrinol.*, **12**, 1441–1451.

Wieland, S., Lan, N.C., Mirasedeghi, S. and Gee, K.W. (1991) Anxiolytic activity of the progesterone metabolite 5 alpha-pregnan-3 alpha-o1-20-one. *Brain Res.*, **565**, 263–268.

Wu, F.S., Gibbs, T.T. and Farb, D.H. (1991) Pregnenolone sulfate: a positive allosteric modulator at the *N*-methyl-D-aspartate receptor. *Mol. Pharmacol.*, **40**, 333–336.

Yamano, S., Aoyama, T., McBride, O.W., Hardwick, J.P., Gelboin, H.V. and Gonzalez, F.J. (1989) Human NADPH-P450 oxidoreductase: complementary DNA cloning, sequence and vaccinia virus-mediated expression and localization of the CYPOR gene to chromosome 7. *Mol. Pharmacol.*, **36**, 83–88.

Yamazaki, M., Mori, H., Araki, K., Mori, K.J. and Mishina, M. (1992) Cloning, expression and modulation of a mouse NMDA receptor subunit. *FEBS Lett.*, **300**, 39–45.

Yanagibashi, K. and Hall, P.F. (1986) Role of electron transport in the regulation of the lyase activity of C21 side-chain cleavage P-450 from porcine adrenal and testicular microsomes. *J. Biol. Chem.*, **261**, 8429–8433.

Yanase, T., Sasano, H., Yubisui, T., Sakai, Y., Takayanagi, R. and Nawata, H. (1998) Immuno-histochemical study of cytochrome b5 in human adrenal gland and in adrenocortical adenomas from patients with Cushing's syndrome. *Endocr. J.*, **45**, 89–95.

Chapter 7

DHEA metabolism in the brain: production and effects of the 7α-hydroxylated derivative

M. Trincal, J. Loeper, D. Pompon, J.J. Hauw and R. Morfin

INTRODUCTION

Both human and murine species carry out the 7α-hydroxylation of dehydro-epiandrosterone (DHEA) and produce 7α-hydroxy-DHEA mainly in liver and brain (Stárka and Kutova, 1962; Šulcová *et al.*, 1968; Doostzadeh and Morfin, 1996) and to a lesser extent in other tissues and organs (Šulcová and Stárka, 1963; Faredin *et al.*, 1969; Couch *et al.*, 1977; Akwa *et al.*, 1992; Morfin and Courchay, 1994). The cytochrome P450 (P450) responsible for 7α-hydroxylation of DHEA appeared as a key to this process (Warner *et al.*, 1989; Akwa *et al.*, 1992; Morfin and Courchay, 1994; Doostzadeh and Morfin, 1996). Recent discovery of this P450 from rat, mouse and human, identified it as *CYP7B1* from its cDNA sequence (Stapleton *et al.*, 1995; Nelson *et al.*, 1996; Wu *et al.*, 1999). Demonstration of its DHEA 7α-hydroxylating potencies when expressed in HepG2 and HeLa cells (Rose *et al.*, 1997; Wu *et al.*, 1999) provided valuable tools for studies of the relationships between cDNA sequences and activity of the resulting CYP 7B1. Production in lymphoid organs and immunoactivating potencies of the 7α-hydroxy-DHEA were demonstrated in mouse (Morfin and Courchay, 1994; Doostzadeh and Morfin, 1996) and led to the concept of a native steroid counteracting the glucocorticoid-mediated immunosuppressive effects (Stárka *et al.*, 1998; Chmielewski *et al.*, 2000; Morfin *et al.*, 2000). Very few data are available on the effects triggered by the 7α-hydroxy-DHEA produced in the brain, but all of the effects described in other tissues and the new data shown in this review imply that brain production of 7α-hyroxy-DHEA may be a key process for the maintenance and the protection of neuronal and nervous structures.

7α-HYDROXYLATION OF DHEA IN MOUSE, RAT AND HUMAN BRAIN

The 7α-hydroxylation of a 3β-hydroxysteroid substrate (5α-androstene-3β,17β-diol) in the rat pituitary gland was first described in the late 70s (Guiraud *et al.*, 1979) and further studies extended the 7α-hydroxylation to 5-androstene-3β,17β-diol and dehydroepiandrosterone (DHEA) (Warner *et al.*, 1989). Use of the neurosteroids DHEA and pregnenolone (PREG) with rat brain microsomes and astrocytes provided evidence for the 7α-hydroxylation of both steroid substrates (Akwa *et al.*,

1992, 1993). Studies in mouse were then initiated and showed that the extent of neurosteroids 7α-hydroxylation in brain was second to that in liver (Morfin and Courchay, 1994; Doostzadeh and Morfin, 1996). In human, the 7α-hydroxylation of DHEA was described in several fetal tissues, in adult liver, mammary gland and skin (Šulcová et al., 1968; Faredin et al., 1969; Björkhem et al., 1972; Couch et al., 1977; Khalil et al., 1993) but, to our knowledge, no report is available for human brain. Most of the studies carried out in murine species determined that a specific cytochrome P450 was involved in 7α-hydroxylation. Both mitochondria and microsomes from mouse brain provided efficient 7α-hydroxylation of PREG and DHEA (Doostzadeh and Morfin, 1996, 1997). Nevertheless, all further studies were carried out with microsomes.

Investigations in mouse treated with dexamethasone showed an 8-fold increase in circulating 7α-hydroxy-DHEA and a 2-fold increase of DHEA 7α-hydroxylation in liver but surprisingly not in brain. In contrast, metyrapone, a known inhibitor of cytochromes P450, decreased significantly the 7α-hydroxylation in both organs (Attal-Khémis et al., 1998). In another report where human adipose stromal cells were cultured in the presence of dexamethasone, activation of DHEA 7α-hydroxylation was proved (Khalil et al., 1994). The question then arose whether the P450 responsible for the 7α-hydroxylation of DHEA was the same in all tissues and organs where such hydroxylation was detected.

INVOLVEMENT OF *CYP7B1* IN BRAIN

A novel cytochrome P450 was identified in rat from hippocampal transcripts and reported (Stapleton et al., 1995). Because of strong similarities in the cDNA sequence with that of P4507a (that carries out 7α-hydroxylation of cholesterol), this new mouse P450 was termed 7b1 (Nelson et al., 1996). Expression of this P450 7b1 in HeLa cells allowed examination of its catalytic activities for the 7α-hydroxylation of PREG and DHEA (Rose et al., 1997). More recently, the human cDNA of P450 7B1 was sequenced from hippocampal transcripts and the 7α-hydroxylation of DHEA was measured after transfection of human kidney 293/T-cells (Wu et al., 1999).

Using the published sequences, we have isolated hippocampal transcripts from the C57BL/6 mouse strain and produced the cDNA of P450 7b1 by RT-PCR. After transfer of the engineered open reading frame to the pYeDP63 multicopy yeast expression vector, the W(hR) yeast strain was chosen for transformation because of its overexpression of the human NADPH-cytochrome P450 reductase (Urban et al., 1993). Yeast and C57BL/6 mouse brain microsomes, containing the expressed P450 7b1, were used for comparative studies of the DHEA and PREG 7α-hydroxylation kinetic parameters (Morfin et al., 2000). The same approach was then used with the human cDNA of P450 7B1 provided by Dr J.Y.L. Chiang and we report now the K_M values for 7α-hydroxylation of PREG and DHEA by the human P450 7B1 expressed in the transformed W(hR) yeast strain (Table 7.1). When compared with K_M values measured in mice and rat under identical conditions for DHEA 7α-hydroxylation, the human K_M was identical with that of C57BL/6 mouse and 7 times lower than that of *Mus musculus* and rat (Table 7.1).

Table 7.1 Kinetic parameters of DHEA and PREG 7α-hydroxylation resulting from incubations of murine brain microsomes and of microsomes of CYP7B1-transformed cells

7α-hydroxylating source	DHEA substrate[a]			PREG substrate[a]		
	K_M^b	K_{cat}^c	V_{max}^d	K_M^b	K_{cat}^c	V_{ma}^d
Mus musculus P450 7b1 in HeLa cells	14	–	303	4.0	–	36
Rat brain microsomes	14	–	322	4.4	–	39
C57BL/6 mouse brain microsomes	1.3	–	4	0.5	–	4
C57BL/6 mouse P450 7b1 in yeast	2	42	–	0.3	11	–
Human P450 7B1 in yeast	2	190	–	8.0	95	–

Notes
[a]Incubations were carried out in 67 mM phosphate buffer (pH 7.4) containing 1 mM EDTA and 1 mM NADPH at 37 °C for 10 min. Radiometabolite separation and measurement were carried out as described previously (Doostzadeh and Morfin 1996); ^{b}μM; ^{c}min^{-1}; ^{d}pmol·min^{-1}·mg^{-1}.

In marked contrast, the human K_M for 7α-hydroxylation of PREG was twice and 16 times larger than that of C57BL/6 mouse and that of *Mus musculus* and rat, respectively. Differences in cDNA sequences of C57BL/6 mouse and of *Mus musculus* were in 9 bases out of 1,524 analyzed. This resulted in eight codon changes with five mutations being silent at the amino acid level and four mutations leading to amino acid changes at positions 265, 278, 432 and 463, while both steroid and heme binding domains remained unchanged (Table 7.2). Further alignment with rat and human sequences showed conservation with minor changes in the steroid and heme binding domains' identity of amino acids 265 and 432, and the conservative M-463-I substitution (Table 7.2). In contrast, S-278-P as well as F-278-P changes may lead to a structural α-helix distortion that could contribute to the

Table 7.2 Differences (bold letters) in amino-acids of CYP7B1 resulting from alignments of Mus musculus and C57BL/6 mouse sequences. Further alignments with rat and human amino acid sequences are given for comparisons

cDNA P450 7B1 species	References	Amino acid #					
		265	278	348–362	432	440–453	463
		Helix G	Helix G	Steroid binding domain		Heme binding domain	Helix L
Mus musculus	Stapleton et al., 1995	**S**	**S**	LESTILEVLRLCSYS	**R**	FGLGTSKCPGRYFA	**E**
C57BL/6 mouse	–		**R**		**K**		**M**
Rat	Stapleton et al., 1995	**R**	**F**	LESTILEVLRLCSYS	**K**	FGLGTSKCPGRYFA	**I**
Human	Wu et al., 1999	**R**	**L**		**K**		**I**
				LES**A**ILEVLRLCSYS		FGLGTSKCPGRYFA	
				LES**SI**FE**A**LRL**S**SYS		FG**T**GTSKCPGR**F**FA	

increased K_M measured in rat and *Mus musculus*. Because production yields of 7α-hydroxylated steroids are strongly dependent from the K_M of P450 7B1, and because the 7α-hydroxylated steroid produced are involved to trigger immunity (Morfin and Courchay, 1994) and may be important for neuroprotection, this paradigm justifies further examination of the human P450 7B1.

ROLE OF P450 7B1 AND SUBSTRATE SPECIFICITY

While cholesterol and oxysterols are native substrates for 7α-hydroxylation by the P450 7A (Ogishima *et al.*, 1987; Norlin *et al.*, 2000), many 3β-hydroxysteroids and even estrogens are 7α-hydroxylated by the P450 7B1 or by P450 7B1-containing nervous tissue preparations (Akwa *et al.*, 1992; Doostzadeh and Morfin, 1996; Rose *et al.*, 1997; Schwarz *et al.*, 1997). The 3β-hydroxysteroid substrates included oxysterols such as 27-, 25- and 24-hydroxy-cholesterol, that, once 7α-hydroxylated, led to the production of primary bile acids (Schwarz *et al.*, 1997; Zhang *et al.*, 1997a) by a second pathway different from classical cholesterol 7α-hydroxylation (Schwarz *et al.*, 1997). Moreover, Zhang *et al.* (1995, 1997b) found with mouse thymocytes and human fibroblasts that 7α-hydroxylation protected cells from the pro-apoptotic effects of oxysterols. Defect in the human P450 7B1 gene was reported to result into severe cholestasis because of accumulation of cytoxic oxysterols (Setchell *et al.*, 1998).

The 7α-hydroxylation of other 3β-hydroxysteroids, including 5α-androstane-3β,17β-diol and 3β-hydroxy-5α-pregnane-20-one, was reported in human, canine and rat prostate and in rat brain (Morfin *et al.*, 1977; Isaacs *et al.*, 1979; Ofner *et al.*, 1979; Strömstedt *et al.*, 1993). Because 3β-hydroxy-5α-pregnane-20-one is a substrate for 7α-hydroxylation and is a native derivative of the anesthetic 3α-hydroxy-5α-pregnane-20-one, it was suggested that 7α-hydroxylation was involved in the regulation of anesthetic steroid levels in brain (Figure 7.1) (Strömstedt *et al.*, 1993). Further support for this paradigm was provided in human and rat prostate where 5α-dihydrotestosterone is the active androgen hormone (Ofner *et al.*, 1979; Morfin, 1988). Thus, 5α-dihydrotestosterone metabolism in prostate provided 5α-androstane-3β,17β-diol which was irreversibly 7α-hydroxylated into androgen inactive derivatives (Sunde *et al.*, 1982; Celotti *et al.*, 1983).

CYP7B1 IN HUMAN HIPPOCAMPUS

Mouse polyclonal antibodies to human P450 7B1 were raised: the full length cDNA of P450 7B1 provided by Dr J.Y.L. Chiang (Wu *et al.*, 1999) was amplified in the *E. coli* DH5-1 strain and provided to *Eurogentec Bel S.A.* (Herstal, Belgium) for DNA immunization of mice. The reactivity of P450 7B1 antibodies in immunsera was detected by Western blotting with human P450 7B1 expressed in yeast microsomes. The antibodies produced were then used for immunohistochemical detection of P450 7B1 on slices (5 µm thickness) of available paraffin-embedded human brain tissue blocks. Peroxidase-bound anti-mouse IgGs were used for detection and the red-brown color was developed by incubation with the DAB

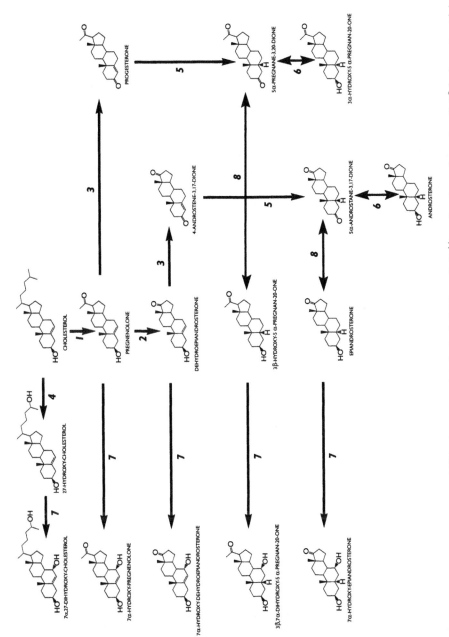

Figure 7.1 Metabolic pathways of 3β-hydroxysteroids in brain. 1. P450scc; 2. Fe^{++}-catalyzed transformation; 3. 3β-hydroxysteroid dehydrogenase; 4. choles-terol 27-hydroxylase; 5. 5α-reductase; 6. 3α-hydroxysteroid oxidoreductase; 7. P4507B1; 8. 3β-hydroxysteroid oxidoreductase.

substrate, and slices were counter-colored with eosin for a pink background and purple nuclei.

Normal human hippocampus remained unstained when mouse pre-immun serum was used (Figure 7.2a). With P450 7B1 immunserum, slices from the same tissue showed a strong P450 7B1-specific staining of neurons while glial cells remained unstained (Figure 7.2b). In contrast, when brain white matter was exposed to the mouse immun serum, no P450 7B1-specific staining was obtained in any cell (Figure 7.2c). Because of the hippocampal origin of P450 7B1 transcripts and resulting cDNA, it was obvious that P450 7B1 had to be found in human hippocampus. Our findings indicate that in human hippocampus the P450 7B1 is exclusively located in pyramidal neurons, and that pathological neuronal deletion could correlate with a decreased P450 7B1 availability and a related decrease in DHEA 7α-hydroxylation.

NEUROPROTECTION AND P450 7B1 ACTIVITY

Neuroprotection may result either from the 7α-hydroxylation of steroids toxic to the nervous cells or from cell protection conferred by the 7α-hydroxylated steroids produced. The first case may be illustrated with both 25- and 27-hydroxycholesterol that are known to induce apoptosis of thymocytes and to be 7α-hydroxylated in mouse thymus and in rat glial cells and neurons (Zhang et al., 1997a,b). After 7α-hydroxylation of these oxysterols, induction of thymocyte apoptosis was abolished. Thus it seems likely that decrease in oxysterols due to 7α-hydroxylation was a key protective event for thymocytes. Whether these oxysterols induce apoptosis in nervous cells and whether neuroprotection results from oxysterol 7α-hydroxylation have not been proved yet. The second case may be illustrated by DHEA and epiandrosterone (EPIA) that did not appear to be toxic to human lymphocytes and to mouse thymocytes (Lafaye et al., 1999; Chmielewski et al., 2000), while their respective 7α-hydroxylated metabolites protected the cells from dexamethasone-induced apoptosis (Chmielewski et al., 2000).

Whether such effect may occur in brain cells has not been investigated. Several works implied glucocorticoids in rat hippocampal neuron loss (Crossin et al., 1997; Joels et al., 1998; Sapolsky et al., 1999; Almeida et al., 2000) and showed that DHEA acted as a neuroprotector through unproved antiglucocorticoid effects (Cardounel et al., 1999; Kimonides et al., 1999; Brown et al., 2000). These data were limited to DHEA and did not consider its transformation into 7α-hydroxy-DHEA by the brain cell preparations. We have shown in rat and mouse that 7α-hydroxy-DHEA was more efficient than DHEA in decreasing the nuclear uptake of the activated gluco-corticoid receptor complex and thus could act as a native antiglucocorticoid (Morfin et al., 1994; Stárka et al., 1998; Morfin and Stárka, 2001). Because of marked differences between murine and human in steroid substrates 7α-hydroxylation (Morfin et al., 2000) and of the absence of glial cell localization of P450 7B1 in human when compared to rat (Akwa et al., 1993; Zhang et al., 1997a), the antiglucocorticoid effects of 7α-hydroxy-DHEA in human could be much more extensive than those described for DHEA (Kalimi et al., 1994). Support for this proposal comes from our finding that in contrast to murine, human lymphocytes of a lymphoid organ and

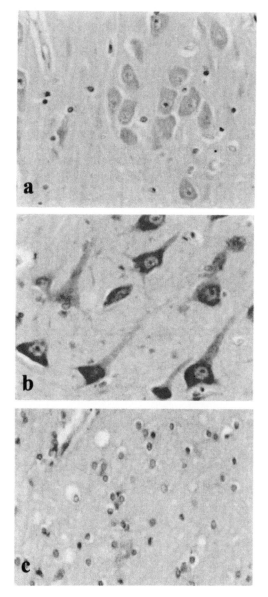

Color plate 1 Immunohistochemical detection of P450 7B1 in human tissue slices. Mouse anti-human P450 7B1 and peroxidase-bound anti-mouse IgGs were used for immunodetection in 5 μm slices obtained from paraffin-embedded tissues. (a) Ammon horn with pre-immun mouse serum (x40). (b) Amon horn with mouse anti-human P450 7B1 (x40). (c) Human brain white matter with mouse anti-human P450 7B1 (x40). Paraffin was removed from slices with xylene and ethanol washes, and P450 7B1 epitopes were unmasked by citrate/citric acid baking in a 400 W microwave oven. Peroxidase inhibition and saturation were carried out with H_2O_2-methanol-water (3:30:67 v/v) before washing and addition of 20% bovine serum albumin, respectively. Incubation with the diluted (1/100) P450 7B1 antibodies was carried out at 4°C for 18h. After extensive washing in tris buffered saline, antibodies bound to P450 7B1 were recognized by anti-mouse IgGs bound to streptavidine-peroxidase (Chenmate Dako kit). (*See page 123*)

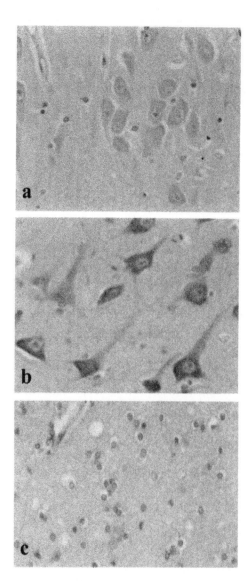

Figure 7.2 Immunohistochemical detection of P450 7B1 in human tissue slices. Mouse anti human P450 7B1 and peroxidase-bound anti-mouse IgGs were used for immuno-detection in 5 μm slices obtained from paraffin-embedded tissues. (a) Ammon horn with pre-immun mouse serum (x40). (b) Amon horn with mouse anti-human P450 7B1 (x40). (c) Human brain white matter with mouse anti-human P450 7B1 (x40). Paraffin was removed from slices with xylene and ethanol washes, and P450 7B1 epitopes were unmasked by citrate/citric acid baking in a 400 W microwave oven. Peroxidase inhibition and saturation were carried out with H_2O_2-methanol-water (3:30:67 v/v) before washing and addition of 20% bovine serum albumin, respectively. Incubation with the diluted (1/100) P450 7B1 antibodies was carried out at 4 °C for 18 h. After extensive washing in tris buffered saline, antibodies bound to P450 7B1 were recognized by anti-mouse IgGs bound to streptavidine-peroxidase (Chenmate Dako kit). (*See Color plate 1*)

blood lymphocytes were unable to carry out 7α-hydroxylation and that both 7α-hydroxy-DHEA and 7α-hydroxy-EPIA triggered the immune response while their steroid precursors were inactive (Lafaye et al., 1999; Chmielewski et al., 2000; Morfin et al., 2000). In the same studies, and because PREG and 7α-hydroxy-PREG were active in mouse and inactive in human, these findings correlate the present demonstration of a K_M higher in human than in mouse or rat for the 7α-hydroxylation of PREG by P450 7B1. Therefore, 7α-hydroxy-DHEA and 7α-hydroxy-EPIA that are readily produced in human tissues could exert neuroprotective effects because of antiglucocorticoid potencies that need to be thoroughly investigated.

CONCLUDING REMARKS

Even though brain, liver and skin contribute to circulating 7α-hydroxy-DHEA, blood levels in human and mouse are in the nM range (Attal-Khemis et al., 1998; Lapcik et al., 1999). In any case, such levels depend both on quantities of available DHEA and on the K_M of P450 7B1. The 7α-hydroxy-DHEA concentrations needed for antiglucocorticoid or immunoactivating effects are 3 orders of magnitude larger than those measured in blood (Stárka et al., 1998; Lafaye et al., 1999; Chmielewski et al., 2000). This led to the concept of a paracrine mode of action of 7α-hydroxy-DHEA towards specific immunoglobulin formation (Lafaye et al., 1999; Morfin and Stárka, 2001).

If the 7α-hydroxy-DHEA produced in neurons induces a neuroprotection, the necessary concentrations will depend strongly on the DHEA supply in brain and on the potencies of P450 7B1 to carry out 7α-hydroxylation in neurons. Circulating DHEA levels decrease with ageing more in male than in female (Šulcová et al., 1997; Hampl et al., in this book) but this sex-associated difference is reversed in old age for 7α-hydroxy-DHEA (Hampl et al., in this book). Thus, it is possible that the onset of brain function diseases associated with old age would take place more in female than in male. In addition, an eventual polymorphism of P450 7B1 in hippocampus could result in a decrease of its catalytic properties and contribute to aggravation of the disease. A recent report of Cascio et al. (2000), shows that in human brain only glial cells may produce DHEA by an alternate reactive oxygen species-mediated pathway and provide it to neurons, and brings indirect support to this suggestion. One may speculate that DHEA produced in human glial cells could be provided to neighboring neurons where 7α-hydroxylation takes place, and that neuroprotection induced by 7α-hydroxy-DHEA is first autocrine in neurons and second paracrine in glial cells. Thus, it is now obvious that interest is arising on 7α-hydroxy-DHEA production and effects, and further investigations should be devoted to deciphering its exact antiglucocorticoid and immunostimulating mode of action and to extensive studies of the human P450 7B1 responsible for its production.

Another hydroxylation product of DHEA hydroxylation is 7β-hydroxy-DHEA (Doostzadeh et al., 1997). The P450 responsible for its production has not yet been identified in any animal or human tissue. It is known that a small portion of the DHEA hydroxylated by P450 7B1 is 7β-hydroxy-DHEA (Rose et al., 1997; Morfin and Stárka, 2001), but there is no way at present to measure accurately the contribution of these quantities to those found in blood (Lapcik et al., 1998, 1999). A possible epimerization of 7α-hydroxy-DHEA through oxidoreduction of 7-oxo-DHEA has been mentioned (Hampl and starka, 1969; Hampl *et al.*, 2000) but such enzyme has not been identified in brain yet. Nevertheless, because of several descriptions of 7β-hydroxy-DHEA mediated increase of immune response and antiglucocorticoid potencies (Loria and Pagett, 1998; Sterzl *et al.*, 1999), this steroid may be implied as much as 7α-hydroxy-DHEA in neuroprotection. In order to answer all key questions related to 7β-hydroxy-DHEA, more investigations need to be carried out on the production and mechanism of action of that steroid.

ACKNOWLEDGMENT

This work was supported in part by a grant from Hunter-Fleming Ltd (Salisbury, UK).

REFERENCES

Akwa, Y., Morfin, R., Robel, P. and Baulieu, E.E. (1992) Neurosteroid metabolism. 7α-Hydroxylation of dehydroepiandrosterone and pregnenolone by rat brain microsomes. *Biochem. J.*, **288**, 959–964.

Akwa, Y., Sananès, N., Gouézou, M., Robel, P., Baulieu, E.E. and Le Goascogne, C. (1993) Astrocytes and neurosteroids: metabolism of pregnenolone and dehydroepiandrosterone. Regulation by cell density. *J. Cell Biol.*, **121**, 135–143.

Attal-Khémis, S., Dalmeyda, V. and Morfin, R. (1998) Change of 7α-hydroxy-dehydroepiandrosterone levels in serum of mice treated by cytochrome P450-modifying agents. *Life Sci.*, **63**, 1543–1553.

Björkhem, I., Einarsson, K., Gustafsson, J.-Å. and Somell, A. (1972) Metabolism of 3β-hydroxy-Δ_5- and 3-oxo-Δ_5-C_{19} and C_{21} steroids in human liver microsomes. *Acta Endocrinol. (Kbh)*, **71**, 569–588.

Brown, R.C., Cascio, C. and Papadopoulos, V. (2000) Pathways of neurosteroid biosynthesis in cell lines from human brain: regulation of dehydroepiandrosterone formation by oxidative stress and beta-amyloid peptide. *J. Neurochem.*, **74**, 847–859.

Cardounel, A., Regelson, W. and Kalimi, M. (1999) Dehydroepiandrosterone protects hippocampal neurons against neurotoxin-induced cell death. *Proc. Soc. Exp. Biol. Med.*, **222**, 145–149.

Celotti, F., Avogadri, N., Ferraboschi, P., Motta, M., Negri-Cesi, P. and Santaniello, E. (1983) Effects of 6- and 7-hydroxy metabolites of 3β,17β-dihydroxy-5α-androstane on gonadotrophin and prolactin secretion and on sex accessories weight of male rats. *J. Steroid Biochem.*, **18**, 397–401.

Chmielewski, V., Drupt, F. and Morfin, R. (2000) Dexamethasone-induced apoptosis of mouse thymocytes: prevention by native 7α-hydroxysteroids. *Immunol. Cell Biol.*, **78**, 238–246.

Couch, R.A.F., Skinner, S.J.M., Tobler, C.J.P. and Doouss, T.W. (1977) The *in vitro* synthesis of 7-hydroxy-dehydroepiandrosterone by human mammary tissue. *Steroids*, **26**, 1–15.

Crossin, K.L., Tai, M.H., Krushel, L.A., Mauro, V.P. and Edelman, G.M. (1997) Glucocorticoid receptor pathways are involved in the inhibition of astrocyte proliferation. *Proc. Natl. Acad. Sci. USA*, **94**, 2687–2692.

Doostzadeh, J., Cotillon, A.C. and Morfin, R. (1997) Dehydroepiandrosterone 7α- and 7β-hydroxylation in mouse brain microsomes. Effects of cytochrome P450 inhibitors and structure-specific inhibition by steroid hormones. *J. Neuroendocrinol.*, **9**, 923–928.

Doostzadeh, J. and Morfin, R. (1996) Studies of the enzyme complex responsible for pregnenolone and dehydroepiandrosterone 7α-hydroxylation in mouse tissues. *Steroids*, **61**, 613–620.

Doostzadeh, J. and Morfin, R. (1997) Effects of cytochrome P450 inhibitors and of steroid hormones on the formation of 7-hydroxylated metabolites of pregnenolone in mouse brain microsomes. *J. Endocrinol.*, **155**, 343–350.

Faredin, I., Fazekas, A.G., Tóth, I., Kokai, K. and Julesz, M. (1969) Transformation *in vitro* of [4-^{14}C]-dehydroepiandrosterone into 7-oxygenated derivatives by normal human male and female skin tissue. *J. Invest. Dermatol.*, **52**, 357–361.

Guiraud, J.M., Morfin, R., Ducouret, B., Samperez, S. and Jouan, P. (1979) Pituitary metabolism of 5α-androstane-3β, 17β-diol: Intense and rapid conversion into 5α-androstane-3β,6α,17β-triol and 5α-androstane-3β,7α,17β-triol. *Steroids*, **34**, 241–248.

Hampl, R., Lapcik, O., Hill, M., Klak, J., Kasal, A., Nováček, A. *et al.* (2000) 7-Hydroxy-dehydroepiandrosterone – a natural antiglucocorticoid and a candidate for steroid replacement therapy? *Physiol. Res.*, **49**, S107–S112.

Hampl, R. and Stárka, L. (1969) Epimerization of naturally occuring C$_{19}$-steroid allyl alcohols by rat liver preparations. *J. Steroid Biochem.*, **1**, 47–56.

Isaacs, J.T., McDermott, I.R. and Coffey, D.S. (1979) Characterization of two enzymic activties of the rat ventral prostate: 5α-androstane-3β,17β-diol, 6α-hydroxylase and 5α-androstane-3β,17β-diol, 7α-hydroxylase. *Steroids*, **33**, 675–692.

Joels, M. and Vreugdenhil, E. (1998) Corticosteroids in the brain – Cellular and molecular actions. *Mol. Neurobiol.*, **17**, 87–108.

Kalimi, M., Shafagoj, Y., Loria, R., Padgett, D. and Regelson, W. (1994) Antiglucocorticoid effects of dehydroepiandrosterone [DHEA]. *Molec. Cell. Biochem.*, **131**, 99–104.

Khalil, M.W., Strutt, B., Vachon, D. and Killinger, D.W. (1993) Metabolism of dehydroepiandrosterone by human adipose stromal cells. Identification of 7α-hydroxy-dehydroepiandrosterone as a major metabolite using high performance liquid chromatography and mass spectrometry. *J. Steroid Biochem. Molec. Biol.*, **46**, 585–595.

Khalil, M.W., Strutt, B., Vachon, D. and Killinger, D.W. (1994) Effect of dexamethasone and cytochrome P450 inhibitors on the formation of 7α-hydroxy-dehydroepiandrosterone by human adipose stromal cells. *J. Steroid Biochem. Molec. Biol.*, **48**, 545–552.

Kimonodes, V.G., Spillantini, M.G., Sofroniew, M.V., Fawcett, J.W. and Hebert, J. (1999) Dehydroepiandrosterone antagonizes the neurotoxic effects of corticosterone and translocation of stress-activated protein kinase 3 in hippocampal primary cultures. *Neuroscience*, **89**, 429–436.

Lafaye, P., Chmielewski, V., Nato, F., Mazié, J.C. and Morfin, R. (1999) The 7α-hydroxy-steroids produced in human tonsils enhance the immune response to tetanus toxoid and *Bordetella pertussis* antigens. *Biochem. Biophys. Acta*, **1472**, 222–231.

Lapcík, O., Hampl, R., Hill, M., Biciková, M. and Stárka, L. (1998) Immunoassay of 7-hydroxysteroids: 1. Radioimmunoassay of 7β-hydroxy-dehydroepiandrosterone. *J. Steroid Biochem. Molec. Biol.*, **67**, 439–445.

Lapcík, O., Hampl, R., Hill, M. and Stárka, L. (1999) Immunoassay of 7-hydroxysteroids: 2. Radioimmunoassay of 7α-hydroxy-dehydroepiandrosterone. *J. Steroid Biochem. Molec. Biol.*, **71**, 231–237.

Loria, R.M. and Pagett, D.A. (1998) Control of the immune response by DHEA and its metabolites. *Rinsho Byori.*, **46**, 505–517.

Morfin, R. (1988) Hydroxylation des stéroïdes androgènes dans leurs organes cibles. *Path. Biol.*, **36**, 925–932.

Morfin, R., Di Stéfano, S., Charles, J.F. and Floch, H.H. (1977) Precursors for 6α- and 7α-hydroxylations of 5α-androstane-3β,17β-diol by human normal and hyperplastic prostates. *Biochimie*, **59**, 637–644.

Morfin, R., Calvez, D. and Malewiak, M.I. (1994) Native immunoactivating steroids interfere with the nuclear binding of glucocorticoids. *9th Intl. Congr. Hormonal Steroids*, Dallas, TX, september 1994. Poster D103.

Morfin, R. and Courchay, G. (1994) Pregnenolone and dehydroepiandrosterone as precursors of native 7-hydroxylated metabolites which increase the immune response in mice. *J. Steroid Biochem. Molec. Biol.*, **50**, 91–100.

Morfin, R., Lafaye, P., Cotillon, A.C., Nato, F., Chmielewski, V. and Pompon, D. (2000) 7α-Hydroxy-dehydroepiandrosterone and immune response. In A. Conti, G.J.M. Maestroni, S.M. McCann, E.M. Sternberg, J.M. Lipton and C.C. Smith (eds), *Neuroimmunomodulation: Perspectives at the new millenium. Ann. N. Y. Acad. Sci.*, **917**, 971–982.

Morfin, R. and Stárka, L. (2001) Neurosteroid 7-hydroxylation products in the brain. In G. Biggio and R.H. Purdy (eds), *Neurosteroids and Brain Function*. Academic Press. San Diego USA. *Int. Rev. Neurobiol.*, **46**, 79–95.

Nelson, D.R., Koymans, L., Kamataki, T., Stegeman, J.J., Feyereisen, R., Waxman, D.J. *et al.* (1996) The P450 superfamily: update on new sequences, gene mapping, accession numbers and nomenclature. *Pharmacogenetics*, **6**, 1–42.

Norlin, N., Andersson, U., Björkhem, I. and Wikvall, K. (2000) Oxysterol 7α-hydroxylase activity by cholesterol 7α-hydroxylase (CYP7A). *J. Biol. Chem.*, **275**, 34046–34053.

Ofner, P., Vena, R.L., Leav, I. and Hamilton, D.W. (1979) Metabolism of C_{19}-radiosteroids by explants of canine prostate and epididymis with disposition as hydroxylated products: A possible mechanism for androgen inactivation. *J. Steroid Biochem.*, **11**, 1367–1379.

Ogishima, T., Degushi, S. and Okuda, K. (1987) Purification and characterization of cholesterol 7α-hydroxylase from rat liver microsomes. *J. Biol. Chem.*, **262**, 7646–7650.

Rose, K.A., Stapleton, G., Dott, K., Kieny, M.P., Best, R., Schwarz, M. *et al.* (1997) Cyp 7b, a novel brain cytochrome P450, catalyzes the synthesis of neurosteroids 7α-hydroxy-dehydroepiandrosterone and 7α-hydroxy-pregnenolone. *Proc. Natl. Acad. Sci. USA*, **94**, 4925–4930.

Sapolsky, R.M. (1999) Glucocorticoids, stress and their adverse neurological effects: relevance to aging. *Exp. Gerontol.*, **34**, 721–732.

Schwarz, M., Lund, E.G., Lathe, R., Björkhem, I. and Russell, D.W. (1997) Identification and characterization of a mouse oxysterol 7α-hydroxylase cDNA. *J. Biol. Chem.*, **272**, 23995–24001.

Setchell, K.D.R., Schwarz, M., O'connell, N., Lund, E.G., Davis, D.L., Lathe, R. *et al.* (1998) Identification of a new inborn error in bile acid synthesis: mutation of the oxysterol 7α-hydroxylase gene causes severe neonatal liver disease. *J. Clin. Invest.*, **102**, 1690–1703.

Sunde, A., Aareskjold, K., Haug, E. and Eik-Nes, K.B. (1982) Synthesis and androgen effects of 7α,17β-dihydroxy-5α-androstan-3-one, 5α-androstan-3β,7α,17β-triol and 5α-androstan-3β,7β,17β-triol. *J. Steroid Biochem.*, **16**, 483–488.

Stapleton, G., Steel, M., Richardson, M., Mason, J.O., Rose, K.A.,. Morris, R.G.M. *et al.* (1995) A novel cytochrome P450 expressed primarily in brain. *J. Biol. Chem.*, **270**, 29739–29745.

Stárka, L. and Kutova, J. (1962) 7-Hydroxylation of dehydroepiandrosterone by rat liver homogenate. *Biochim. Biophys. Acta*, **56**, 76–82.

Stárka, L., Hill, M., Hampl, R., Malewiak, M.I., Benalycherif, A., Morfin, R. *et al.* (1998) Studies on the mechanism of antiglucocorticoid action of 7α-hydroxy-dehydroepiandrosterone. *Collect. Czech. Chem. Commun.*, **63**, 1683–1698.

Šterzl, I., Hampl, R., Šterzl, J., Votruba, J. and Stárka, L. (1999) 7β-Hydroxy-DHEA counteracts dexamethasone induced suppression of primary immune response in murine splenocytes. *J. Steroid Biochem. Molec. Biol.*, **71**, 133–137.

Strömstedt, M., Warner, M., Banner, C.D., Macdonald, P.C. and Gustafsson, J.-Å. (1993) Role of brain cytochrome P450 in regulation of the level of anesthetic steroids in brain. *Molec. Pharmacol.*, **44**, 1077–1083.

Šulcová, J. and Stárka, L. (1963) Extrahepatic 7α-hydroxylation of dehydroepiandrosterone. *Experimentia*, **19**, 632–633.

Šulcová, J., Capkova, A., Jirasek, J.E.V. and Stárka, L. (1968) 7-Hydroxylation of dehydroepiandrosterone in human fetal liver, adrenals and chorion *in vitro*. *Acta Endocr. (Kbh)*, **59**, 1–9.

Šulcova, J., Hill, M., Hampl, R. and Stárka, L. (1997) Age and sex related differences in serum levels of unconjugated dehydroepiandrosterone and its sulphate in normal subjects. *J. Endocrinol.*, **154**, 57–62.

Urban, P., Truan, G., Gautier, J.C. and Pompon, D. (1993) Xenobiotic metabolism in humanized yeast: engineered yeast cells producing human NADPH-cytochrome P450 reductase, cytochrome b_5, epoxide hydrolase and P-450s. *Biochem. Soc. Trans.*, **21**, 1028–1034.

Warner, M., Tollet, P., Strömstedt, M., Carlström, K. and Gustafsson, J.-Å. (1989) Endocrine regulation of cytochrome P-450 in the rat brain and pituitary gland. *J. Endocrinol.*, **122**, 341–349.

Wu, Z., Martin, K.O., Javitt, N.B. and Chiang, J.Y.L. (1999) Structure and functions of human oxysterol 7α-hydroxylase cDNAs and gene *CYP7B1*. *J. Lipid Res.*, **40**, 2195–2203.

Zhang, J., Larsson, O. and Sjövall, J. (1995) 7α-Hydroxylation of 25-hydroxycholesterol and 27-hydroxy-cholesterol in human fibroblasts. *Biochim. Biophys. Acta*, **1256**, 353–359.

Zhang, J., Akwa, Y., El-Etr, M., Baulieu, E.E. and Sjövall, J. (1997a) Metabolism of 27-, 25- and 24-hydroxycholesterol in rat glial cells and neurons. *Biochem. J.*, **322**, 175–184.

Zhang, J., Xue, Y., Jondal, M. and Sjövall, J. (1997b) 7α-Hydroxylation and 3-dehydrogenation abolish the ability of 25-hydroxycholesterol and 27-hydroxycholesterol to induce apoptosis in thymocytes. *Eur. J. Biochem.*, **247**, 129–135.

Chapter 8

DHEA: biosynthesis, regulation and function in the central nervous system

Rachel C. Brown, Ying Liu and Vassilios Papadopoulos

INTRODUCTION

It has been now well established that steroid hormones act by regulating gene expression. This action of steroids is a classic mechanism of inducing cellular processes such as growth and differentiation in steroid-sensitive tissues. The genomic effects of steroids are mediated through proteins that are members of the superfamily of steroid hormone receptors, a group of intracellular transcription factors (Beato, 1989). These receptors reside within the cell in an inactive form until binding to their steroid ligand, which causes a conformation change in the receptor, allowing it to bind to DNA (Figure 8.1). At this point, the steroid–receptor complex is translocated to the nucleus where it binds to various steroid response elements on the promoters of a number of genes, and alters gene expression (reviewed in Miesfield, 1989). These genomic actions of steroids are characterized by occurring over a long period of time, ranging from many hours to days to years. The steroid hormone receptor superfamily includes the glucocorticoid and mineralocorticoid receptors for stress steroids, progesterone, estrogen and androgen receptors for the sex steroids, thyroid hormone, vitamin D and retinoic acid receptors.

Steroids also have rapid, non-genomic effects, particularly in the brain, that were first observed in the 1940s (Selye, 1941, 1942). These actions initially involved anesthetic metabolites of progesterone, but have since been expanded to include a large number of steroid compounds. The non-genomic activity of steroids is characterized by extremely rapid effects, lasting from milliseconds to minutes, and does not require interaction with steroid hormone receptors (Orchinik and McEwe, 1993). In the central nervous system (CNS), these effects are thought to involve steroid modulation of membrane-bound neurotransmitter receptors (Majewska, 1987; Lambert *et al.*, 1995), including the GABA$_A$ receptor complex and the NMDA class of glutamate receptors (Mellon, 1994, Figure 8.2). Neuroactive steroids have been intensively studied in recent years, due to their great appeal as potential drugs for treatment of a number of neuropathological and clinical conditions. Because of their lipophilic structure, steroids can easily diffuse across the blood–brain barrier if given peripherally, thereby bypassing the issues of drug delivery from the circulation across the blood–brain barrier and into the brain. Furthermore, the amounts of steroid needed to induce changes in neural activity are extremely low, typically in the

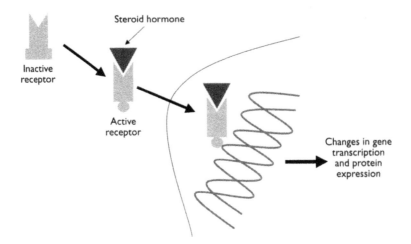

Figure 8.1 Classical action of steroid hormones. The classic action of steroid hormones is due to the presence of intracellular steroid hormone receptors, which exist in the cytoplasm in an inactive conformation. When these receptors bind their steroid ligands, they undergo a conformational change, leading to activation. The receptor–steroid complexes then translocate to the nucleus, where they bind to sterol response elements in the genome and alter gene transcription. These effects are slowly occurring and long lasting, in contrast to the non-genomic effects of steroids, which occur over a very short period of time.

nanomolar range. The steroids that have attracted the greatest amount of interest in this arena are the 5α-reduced metabolites of progesterone, and a steroid called dehydroepiandrosterone, or DHEA. The rest of this review will focus on the biosynthesis and activity of DHEA in the CNS as compared to the periphery (outside the CNS).

DHEA

DHEA is a naturally occurring steroid produced by the adrenal cortex in humans and is the most abundant steroid in the blood under normal conditions (Kroboth *et al.*, 1999). In adult humans, most DHEA secreted by the adrenal glands is released as the sulfated form (DHEAS). Together, DHEA and DHEAS are the precursors for ~50% of androgens produced in adult men, ~75% of estrogen in premenopausal women, and 100% of estrogen produced in postmenopausal women. Under normal conditions, DHEA is secreted with cortisol in response to ACTH stimulation of the adrenal cortex. There is some debate as to whether the function of DHEA is solely as a precursor for androgen biosynthesis or if it has some biological activity of its own. Because there is no known intracellular steroid receptor for DHEA it seems likely that the majority of its actions are due to its conversion to estrogen and testosterone in target tissues. Importantly,

NAS-producing cell

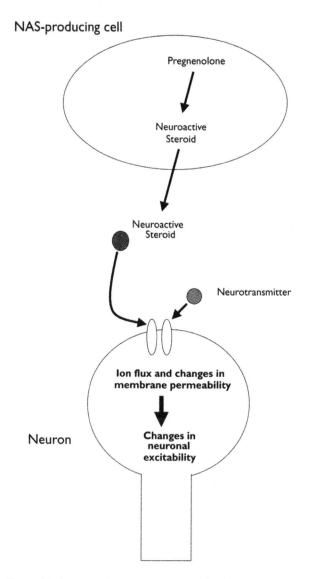

Figure 8.2 Actions of neuroactive steroids. Neuroactive steroids (NAS) can be produced locally within the central nervous system. All NAS are made from pregnenolone, and can leave their sites of production by diffusing through the cell membrane. These steroids can bind to specific sites on a number of neurotransmitter receptors and ion channels, altering ion permeability and thereby modulating neuronal excitability.

DHEA production is age-dependent, with adrenal production peaking at about age 25 and declining to 15–20% of peak levels by age 80 (Baulieu and Robel, 1998).

In recent years, much interest has been directed towards DHEA in the brain and the role of DHEA in ageing. DHEA was first isolated in the brain in 1981

(Corpechot *et al.*, 1981), and has since been demonstrated to be present in the brain at higher levels than in the periphery (Kroboth *et al.*, 1999). Furthermore, brain DHEA levels seem to be independent of peripheral steroid producing tissues (Corpechot *et al.*, 1981, 1983, 1993; Robel *et al.*, 1995), and will persist for up to two weeks following the removal of the adrenal glands and gonads in rats. This suggests either a sequestration of DHEA in the brain, or *de novo* synthesis of DHEA within the brain itself.

BIOSYNTHESIS OF DHEA

Biosynthesis of DHEA in the periphery

All steroid hormones are derived from cholesterol (Figure 8.3). The initial step in steroid biosynthesis is the metabolism of cholesterol to pregnenolone, a process that occurs on the inner mitochondrial membrane, and is catalyzed by the cytochrome P450 side chain cleavage enzyme (P450scc). The rate-limiting step in steroid biosynthesis is the transport of cholesterol from the cytosol to the inner mitochondrial membrane, a process that is mediated by a complex of proteins on the cytoplasmic side of the outer mitochondrial membrane (Papadopoulos, 1993; Brown and Papadopoulos, 2001). DHEA is synthesized in the microsomal compartment of cells of adrenal cortex (as well as other steroidogenic cell types) by the metabolism of pregnenolone via the cytochrome P45017α-hydroxylase/ 17,20-lyase (P450c17). The expression of the P450c17 enzyme mRNA is regulated (increased) by ACTH and its second messenger cAMP (Whitlock, 1986). Prostaglandin E2 can also induce P450c17 mRNA expression (Rainey *et al.*, 1993). Further PCR analysis has demonstrated two products amplified from rat adrenal glands and brain, but only one product in testis (Sanne and Krueger, 1995), suggesting that alternative, tissue-specific, splice variants of the P450c17 enzyme exist. P450c17 is synthesized as a mature protein (Whitlock, 1986). The enzyme activity can be induced by pituitary hormones and cAMP (Whitlock, 1986) and although the enzyme is active on its own, high levels of 17,20-lyase activity require the presence of cytochrome P450 oxidoreductase as an electron donor (Miller *et al.*, 1997). Cytochrome b_5 may also act as an electron donor for P450c17. In the non-human primate adrenal gland, P450c17 is expressed in the zona fasciculata, where it is involved in the production of glucocorticoids, and in the zona reticularis, where it mediates androgen production (Mesiano *et al.*, 1993). In the human adrenal gland, DHEA is also present in the zona reticularis (Miller *et al.*, 1997; Parker 1999). The development of compounds affecting P450c17 activity allowed to better examine the role of the enzyme in *in vitro* and *in vivo* settings. P450c17 activity can be inhibited by SU 10603, a pyridine derivative (Lyne *et al.*, 1974; LaCagnin *et al.*, 1989), the imidazole fungicide ketoconazole (Albertson *et al.*, 1988a,b) or inactivated by 17β-(cyclopropylamino)-androst-5-ene-3β-ol (Murray and Reidy, 1990).

In the rat, where the majority of studies on DHEA synthesis and its neuroactive effects have been done, P450c17 activity and protein expression are limited to the theca cells of the ovary and the Leydig cells of the testis, both primary sources of

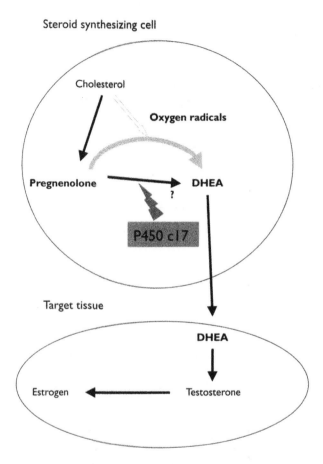

Steroid synthesizing cell

Figure 8.3 Biosynthesis of DHEA. In the adrenal gland, DHEA is synthesized from pregnenolone by the actions of the P450c17 enzyme. DHEA can then be secreted into
the circulation, where it is bound to serum proteins and can travel to distant target
tissues. At these target tissues, DHEA can be converted to testosterone and
estrogen, both of which are known to activate gene transcription. In the brain,
DHEA biosynthesis occurs via the known peripheral enzymatic pathway, and by a
brain-specific alternative pathway involving oxygen radicals. In neurons, DHEA is
synthesized by P450c17 only. However, in glial cells, P450c17 does not seem to
play a large role in DHEA synthesis. These cells can produce DHEA in response to
increases in intracellular free radicals.

estrogen and testosterone respectively (Hall, 1984; Le Goascogne *et al.*, 1991).
Although P450c17 mRNA is present in the rat adrenal (Stromstedt and Waterman, 1995) and embryonic rat central nervous system (CNS) (Compagnone *et al.*,
1995) no P450c17 immunoreactivity and bioactivity were detected in either the
adrenal cortex (Hall, 1984) or the neonatal and adult rat brain (LeGoascogne *et al.*,
1991; Cascio *et al.*, 2001). These data indicate that the presence of an mRNA
transcript does not necessarily reflect the expression of a biologically active

protein. This may explain many studies showing that rats do not make high levels of peripheral DHEA (Hall, 1985), and is intriguing in light of the high levels of DHEA present in rat brain (Baulieu and Robel, 1998).

DHEA and DHEAS are produced in the human, starting at adrenarche (Parker, 1999). In the blood, DHEA and DHEAS are bound to albumin, and DHEA production is extremely sensitive to ACTH stimulation. With age, adrenal production of DHEA slows in both human and non-human primates (Parker, 1999). Serum levels in elderly adults reach only 10–20% of those in young adults. There is also a decrease in DHEA produced in response to stimulation. However, the specific mechanism responsible for this decreased production is still unclear.

Biosynthesis of DHEA in the central nervous system

The synthesis of steroids *de novo* within the CNS has been an active field of research for more than 15 years. An issue in characterizing the physiological role of neurosteroids has been determining their source. Two possibilities have been considered: diffusion of steroids made in peripheral steroidogenic tissues, such as the gonads and adrenals, across the blood–brain barrier, or *de novo* steroid biosynthesis within the brain itself. Baulieu and colleagues have demonstrated that the levels of neurosteroids in the rat brain are very high, and persist for up to two weeks after the removal of peripheral steroidogenic organs (Corpechot *et al.*, 1981, 1983, 1993; Robel *et al.*, 1995). This group has established immunocytochemical localization of the cytochrome P450 side chain cleavage enzyme (P450scc) to the white matter of the rat brain, suggesting that oligodendrocytes are a source of neurosteroids in the brain (LeGoascogne *et al.*, 1987). These pioneering studies have also demonstrated the conversion of pregnenolone to progesterone in rat glial cultures (Hu *et al.*, 1987, 1989; LeGoascogne *et al.*, 1987; Jung-Testas *et al.*, 1989). Other groups have demonstrated the production of progesterone and 5α-reduced metabolites of progesterone by rat brain (Melcangi *et al.*, 1994), which have been shown to be positive allosteric modulators of the GABA$_A$ receptor complex (Majewska *et al.*, 1986; Lambert *et al.*, 1995). There have been a number of studies to further examine the expression and activity of components of the peripheral steroidogenic pathway in the brain. In peripheral steroid biosynthesis, cholesterol is transported from intracellular stores across the outer mitochondrial membrane to the inner mitochondrial membrane, where P450scc is located, through a complex ivolving the peripheral-type benzodiazepine receptor (PBR) (Papadopoulos, 1993; Papadopoulos *et al.*, 1997) and the steroidogenic acute regulatory protein (StAR) (Stocco and Clark, 1996). The P450scc converts cholesterol to pregnenolone. Primary cultures of mixed glia (Hu *et al.*, 1987, 1989; Jung-Testas *et al.*, 1989) metabolize cholesterol to pregnenolone and progesterone. Furthermore, mRNA and protein for the P450scc and 3β-hydroxysteroid dehydrogenase (3β-HSD, converts pregnenolone to progesterone), as well as further progesterone metabolizing enzymes, have been found in specific brain areas and cell types (Papadopoulos *et al.*, 1992; Mellon and Deschepper, 1993; Melcangi *et al.*, 1994; Compagnone *et al.*, 1995; Guennoun *et al.*, 1995). However, convincing evidence for P450c17 expression and activity in the brain has been more difficult to find.

Evidence for P450c17 message, protein and activity in the rat

Despite these observations of peripheral steroidogenic enzymes in the CNS, which indicate that steroids in brain may be formed via the same enzymatic pathways as those described in adrenals and gonads, there is controversy about the mechanism of DHEA synthesis. DHEA, the first neurosteroid described (Corpechot *et al.*, 1981) is a major neuroactive steroid, and constitutes a main portion of the neurosteroids found in the rat brain (Baulieu and Robel, 1998). However, there has been no convincing demonstration of cytochrome P450c17 activity and protein concurrent with P450c17 mRNA expression in the brain (Figure 8.3). Conflicting reports exist on the presence of P450c17 mRNA transcripts in the CNS. Although P450c17 mRNA has been found by *in situ* hybridization to be present during rat embryonic development (Compagnone *et al.*, 1995), there is no solid evidence for P450c17 mRNA in the adult (Mellon and Deschepper, 1993; Stromstedt and Waterman, 1995; Kohchi *et al.*, 1998). Moreover, there has been no demonstration on the presence of either P450c17 protein expression (LeGoascogne *et al.*, 1991; Cascio *et al.*, 2001) or enzyme activity (Akwa *et al.*, 1991; Cascio *et al.*, 2001). As we noted earlier a similar contradiction has been shown in the rat adrenal a tissue devoid of P450c17 activity (Hall, 1984) where the P450c17 mRNA is present (Stromstedt and Waterman, 1995).

These studies leave us with a crucial question – how is DHEA made in the brain if the P450c17 enzyme is absent or inactive? DHEA is the most prevalent neurosteroid in the rat brain (Baulieu and Robel, 1998), and the rat peripheral steroidogenic organs do not make DHEA, so the search for P450c17 has focused largely in the brain in this animal model. A recent study has demonstrated DHEA production when cultures of isolated rat glial cells are incubated with pregnenolone as a precursor (Zwain and Yen, 1999). However, methodological concerns render these data somewhat questionable (Cascio *et al.*, 2001). We are left with the question of how rat brain produces DHEA.

Possibility of an alternative pathway in rat brain

In 1994, Prasad *et al.*, building on the issues first addressed by Lieberman (Lieberman and Prasad, 1990), published a paper in which they reported production of DHEA in organic extracts of rat brain treated with different oxidizing and reducing agents. This paper proposed the existence of alternative precursors for neurosteroids in the brain that could be metabolized under appropriate oxidative conditions. This was the first indication that there could be a novel and specific mechanism for steroid biosynthesis in the brain. Since the Prasad *et al.* (1994) study examined this alternative pathway for DHEA synthesis in an artificial and non-physiological system, we have looked for this alternative pathway in living cells. Studies in our laboratory using Fe^{++} ions as a redox tool have clearly demonstrated the presence of an alternative pathway for DHEA synthesis in rat C6–2B glioma cells, which lack both the P450c17 mRNA (Cascio *et al.*, 1998) and protein (Cascio *et al.*, 1999). This pathway was further localized in the microsomal fraction of the C6–2B glioma cells (Cascio *et al.*, 1998). These findings raise the possibility that brain-derived DHEA may not be made according to the known peripheral

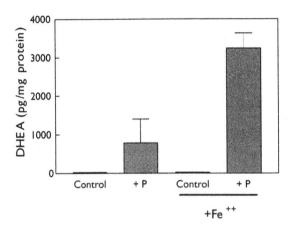

Figure 8.4 DHEA formation by bovine brain microsomes. Bovine brain microsomes incubated in the presence of the substrate pregnenolone (P, 50 μM) and inhibitors of P450c17, SU-10603 (5 μM), and 3β-hydroxysteroid dehydrogenase, Trilostane (5 μM). Microsom\es were treated with or without 10 mM $FeSO_4$ (Fe^{++}) DHEA was extracted, isolated and measured as described by Cascio *et al.* (1998). Data shown is means ±S.D. from an experiment performed in triplicates. Similar results were obtained in two other separate experiments.

enzymatic pathway, but by this alternative, as yet undefined process. The probable alternative precursor being acted upon by this alternative pathway is a 17-hydroperoxide of pregnenolone, which can be converted to DHEA through a process of ketone formation by β-fragmentation (Cascio *et al.*, 1998), although the conversion of a hydroperoxide of cholesterol to DHEA cannot be yet excluded. This alternative pathway was also found in microsomes isolated from neonatal rat brain cortex and primary cultures of rat glial cells (Cascio *et al.*, 2001).

In contrast to the C6–2B glioma cells, P450c17 mRNA and protein were found in immature oligodendrocyte precursors and mature oligodendrocytes. In the presence of substrate (pregnenolone), these cells made DHEA, and addition of Fe^{++} increased DHEA formation in these cells. However, DHEA formation in the presence of pregnenolone and Fe^{++} was not inhibited by the P450c17 inhibitors SU-10603 (Cascio *et al.*, 2000) and ketoconazole (unpublished data), indicating that the formation of DHEA in oligodendrocytes occurs independently of the P450c17 protein present in the cells. Isolated type I astrocytes do not express P450c17 mRNA or protein, but will respond to Fe^{++} by producing DHEA (Cascio *et al.*, 2000). Thus, DHEA production in both types of cells occurred independently of active P450c17. Furthermore, these results suggest that in differentiating oligodendrocytes and astrocytes, DHEA is formed via an oxidative stress-dependent alternative pathway (Figure 8.3).

This alternative pathway is not limited to rat brain and glial cells. Isolated bovine brain microsomes exhibited a similar activity in response to Fe^{++} treatment (Figure 8.4). In these experiments, DHEA formation was examined in the presence of the inhibitor of 3β-hydroxysteroid dehydrogenase trilostane and the inhibitor of P450c17 SU-10603, or ketoconazole, as well as in the presence or absence of the substrate pregnenolone. Under these conditions, addition of Fe^{++} induced DHEA formation in the presence of the substrate pregnenolone. Human cells will also make DHEA via this pathway in a cell-type specific manner. Human oligodendrocytes and astrocytes will make DHEA in response to Fe^{++} treatment, but human neurons will not (Brown *et al.*, 2000a). Moreover, this DHEA formation is regulated by levels of intracellular free radicals and can be blocked by antioxidants such as Vitamin E (Brown *et al.*, 2000a). Furthermore, the activity of the alternative pathway can be regulated by other treatments that cause increases in intracellular free radicals, such as treatment with β-amyloid, a toxic peptide that is a component of Alzheimer's disease neuritic plaques.

Evidence for P450c17 message, protein and activity in human

Despite the prevalent use of various neurosteroids, such as DHEA, as supplements with the goal of improving brain function, specifically age-related problems, there is no information about neurosteroid levels in the human brain under various physiological and pathological conditions. The majority of the studies on neurosteroid biosynthesis have been done in rodents. Although several studies have looked at the regional distribution of neurosteroids in postmortem human brain (Lanthier and Patwardhan, 1986; Lacroix *et al.*, 1987; Bixo *et al.*, 1997), there in no evidence as to whether these steroids were synthesized in the brain or accumulated in specific brain areas from peripheral sources. In addition, there have been no published studies examining mechanisms of regulation of neurosteroid biosynthesis, if it occurs, in either rodents or humans. We have addressed the question of DHEA biosynthesis in the human brain by looking at an *in vitro* cell culture system, and samples of human brain tissue taken from Alzheimer's disease patients and age-matched, non-demented controls.

Our first step was to determine the effect of SU 10603, the specific inhibitor of P450c17 activity, on endogenous levels of DHEA in human glioma cells expressing message for MBP, normal human astrocytes expressing message for glial fibrillary acidic protein (GFAP), an astrocyte marker, and differentiated, post-mitotic human neurons derived from the NTera2 teratocarcinoma cell line by treatment with retinoic acid. 5 µM SU 10603 has no effect on endogenous levels of DHEA from either oligodendrocytes or astrocytes, indicating that DHEA production in these cells is not dependent of P450c17 activity. However, in human neurons, SU 10603 treatment wipes out DHEA formation completely (Figure 8.5), indicating that in neurons, all DHEA formed is due to P450c17 activity. This is the first indication of P450c17 in human brain, and specifically in neurons. However, the fact that SU 10603 does not inhibit glial DHEA production led us to examine these cells further for the presence of the alternative pathway already described in rat glial cells.

Alternative pathway activity in human glial cells

We examined alternative pathway activity in human glial cells by treating the cells with $FeSO_4$ and looking at DHEA production. We found that both oligodendrocytes

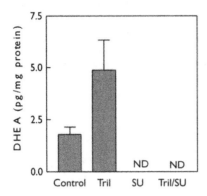

Figure 8.5 DHEA synthesis in human neurons is dependent on P450c17 activity. Normal human neurons were grown in culture, incubated with inhibitors of peripheral steroid bio-synthesis and levels of DHEA were measured. Untreated (control) cells made about 2.1 pg/mg protein. When cells were treated with 10 µM Trilostane (Tril), an inhibitor of progesterone synthesis which increases the pool of pregnenolone that can be converted into DHEA, the levels of DHEA go up to about 5.0 pg/mg protein. However, when the cells are treated with 5 µM SU 10603 (SU), a specific inhibitor of P450c17 activity and DHEA biosynthesis, either alone or in conjunction with Trilostane, no DHEA is detectable by our methods. This indicates the neuronal DHEA synthesis is dependent on P450c17 activity. Furthermore, neurons do not exhibit any alternative pathway activity when treated with appropriate agents.

and astrocytes will produce DHEA in response to $FeSO_4$ (Brown *et al.*, 2000a). This pathway does not appear to be an artifact of cell culture, because treating homogenates of human brain with $FeSO_4$ also results in DHEA production (Brown *et al.*, 2000b). This alternative pathway activity is dependent on a heat and acid sensitive component that we are in the process of characterizing.

Function of the alternative pathway

The mechanism by which we detect alternative pathway activity, treatment with $FeSO_4$, is a very harsh and non-physiologic environment, with high levels of oxidative stress. This has brought up the question of whether or not this pathway is relevant *in vivo*. Recent studies in our laboratory have addressed this question in the case of Alzheimer's disease (AD), by treating cells in culture with β-amyloid (Aβ) and examining samples of human AD brain for evidence of the alternative pathway.

We examined the ability of Aβ, a peptide involved in the pathogenesis of AD that can cause increases in reactive oxygen species and oxidative stress in neurons (Subbarao *et al.*, 1990; Mecocci *et al.*, 1994; Nunomura *et al.*, 1999) to activate the alternative pathway in our cell culture model system. We found that Aβ will increase levels of cellular reactive oxygen species and DHEA after 24 hours of treatment. Both of these increases can be blocked by co-treatment with Vitamin E, an anti-oxidant (Brown *et al.*, 2000a). This suggests that the alternative pathway for DHEA synthesis could be important in pathological conditions involving increased oxidative stress. Furthermore, in tissue samples isolated from the hippocampus, hypothalamus and frontal cortex of severe AD patients, levels of DHEA are significantly increased as compared to age-matched controls. We believe this to be representative of the increased oxidative stress and inflammation thought to be important players in the pathogenesis of AD (Marksberry, 1997).

DHEA IN NORMAL AND PATHOLOGICAL BRAIN FUNCTION

A great deal of attention has been directed towards DHEA as a cure for ageing and age-related pathologies (Baulieu, 1996). We will now address a number of the claims made about DHEA's restorative effects.

Normal actions of DHEA in the brain

What is the normal role of DHEA in the brain? DHEA is considered to be a neuro-active steroid, and as such, can modulate neuronal excitability through a number of different mechanisms (Baulieu and Robel, 1996). DHEA has been shown to enhance memory in male mice (Roberts *et al.*, 1987; Flood *et al.*, 1992) and long-term potentiation in the hippocampus (Yoo *et al.*, 1996). DHEA can potentiate neuronal NMDA responses via sigma receptors (Bergeron *et al.*, 1996), which may account for some of its memory-enhancing effects. There is also some evidence that DHEA may be important in differentiation of glia and neurons during development (Roberts *et al.*, 1987) and in cortical organization (Compagnone and Mellon, 1998). Ultimately the role of DHEA in the brain is, at this time, still unknown. It remains to be seen whether knockout of the P450c17 gene in mice will dramatically affect brain function. Considering that there are no mental, memory or other neurological problems reported in patients exhibiting P450c17-deficiency, it is possible that brain P450c17 may not have any physiological role in neurosteroid biosynthesis.

DHEA and emotional state

Recent studies have examined levels of DHEA and DHEA sulfate as potential con-tributors to depression. Depressed patients have been shown to have significantly higher plasma levels of DHEAS than non-depressed controls (Takebayashi *et al.*, 1998). Treatment with antidepressants decreased DHEA-sulfate levels. Other neuroactive steroids are affected by antidepressant treatment (Romeo *et al.*, 1998). Levels of several neuroactive steroids, primarily the 5α-reduced metabolites of progesterone, are decreased in depression. Treatment with Fluoxetine (Prozac™) restores the levels of these steroids to those of non-depressed patients, and allevi-ates some of their symptoms.

DHEA has also been used as a potential treatment for depression. In 22 patients with 90 mg DHEA/day or placebo, DHEA was associated with a significantly greater decrease in depression rating (Wolkowitz *et al.*, 1999). In another study, 17 men and women with depression were treated for 6 weeks with DHEA or placebo.

DHEA treatment significantly improved scores on several scales of depression, as well as improving energy levels and motivation, and decreasing worry and stress (Bloch et al., 1999). DHEA may be positively involved in mediating depression and mood, although there is no evidence that these effects are not due to the conversion of DHEA to other steroid products. Furthermore, reducing DHEA-sulfate levels is associated with positive changes in mood and perceived stress in HIV+ men (Cruess et al., 1999), and DHEA treatment (50 mg/day for four months) increases reports of overall well-being, and improves depression and anxiety (Arlt et al., 1999). Longer treatments have similar effects (Baulieu et al., 2000).

Several studies have shown that DHEA treatment can improve libido and sexual function probably mediated by the central effects of the steroid. 50 mg/day of DHEA for one year increased libido in women (Baulieu et al., 2000). Another study showed that 50 mg/day of DHEA for four months resulted in increased sexuality/libido in women with adrenal insufficiency (Arlt et al., 1999). DHEA may also be a possible treatment for impotence in men. Low DHEA levels are correlated with erectile dysfunction (Reiter et al., 2000), and 50 mg/day of DHEA for 6 months decreases impotency, but has no effect on markers of prostate health, such as PSA (Reiter et al., 1999).

DHEA and cognitive function: can DHEA prevent AD?

There has been a great deal of research trying to link DHEA with cognitive function, for a number of reasons. Age-related decrease in DHEA is one of the most striking changes in endocrine function over one's lifetime (Parker, 1999). DHEA, as well as other neurosteroids, may be involved in modulating learning and memory (Roberts et al., 1987; Flood et al., 1992; Yoo et al., 1996). However, there is no convincing evidence that DHEA is involved in human age-related cognitive decline or dementia (Guazzo et al., 1996; Wolf et al., 1997a,b). Furthermore, there is no convincing connection between low serum DHEA levels and the development of future AD (Berr et al., 1996).

AD is the most common cause of dementia in people over the age of 65 (1999 Progress Report on Alzheimer's Disease, National Institute of Ageing, National Institutes of Health, Bethesda, MD). NIA estimates that there are 360,000 new cases reported each year and predict that this number will increase dramatically as the population ages. Recent evidence suggests that inflammation and increased oxidative stress may play a large role in the pathogenesis of AD (Markesbery, 1997). β-amyloid is a peptide important in the pathogenesis of AD, and it has been shown to cause increases in free radicals in neurons (Subbarao et al., 1990; Mecocci et al., 1994; Nunomura et al., 1999) and glia (Brown et al., 2000a), directly producing hydrogen peroxide through metal ion reduction (Huang et al., 1999). This suggests that Aβ may play a direct role in increasing oxidative stress in the AD brain.

Recent studies in our laboratory have demonstrated that, although peripheral DHEA levels between AD and age-matched control patients do not differ significantly, there is a significantly higher amount of DHEA present in both the brain and CSF of AD patients (Brown et al., 2000b). We hypothesize that this increase in DHEA is due to increased levels of oxidative stress in the AD brain, possibly

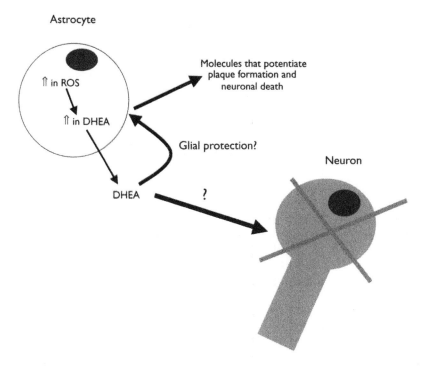

Figure 8.6 A possible role for DHEA in AD: potentiating plaque formation and neurotoxicity. In response to increased oxidative stress, potentially due to Aβ, glia start to make DHEA via the alternative pathway. This increased DHEA production may or may not make it to neurons to protect them against the toxic effects of Aβ. It is likely that the glial DHEA production may protect the glial themselves, and particularly reactive astrocytes. This astrocyte protection may lead to enhancement of the reactive astrocyte response, increasing the production of molecules, like extracellular matrix components, that help to potentiate Aβ deposition and plaque formation. This may enhance the neurotoxic effects of Aβ, leading to increased neuronal death.

mediated by increased levels of toxic aggregated Aβ. We know that Aβ can cause neurons and glia to produce very high levels of free radicals; these oxygen radicals may contribute to neuronal death. There is evidence that DHEA can actually protect neurons against different kinds of toxic insults, including Aβ mediated neurotoxicity (Kimonides *et al.*, 1998; Cardounel *et al.*, 2000). However, we have shown that production of DHEA mediated by oxidative stress occurs only in human glial cells (Brown *et al.*, 2000a). The DHEA made by glia may protect oligodendrocytes and astrocytes, but may not diffuse far enough away from these cells to reach neurons in order to protect them (Figure 8.6). In fact, it may be that DHEA may act solely to protect glia, an event that could potentiate the neuropathology of AD by protecting reactive astrocytes and promoting the productions of extracellular matrix molecules and growth factors that may further potentiate plaque formation in the AD brain (Canning *et al.*, 1993). It remains to be seen if this is indeed the case.

CONCLUSION

The role of DHEA in the brain, as well as in various pathological conditions, is potentially very important. DHEA can modulate neurotransmitter receptor activity, is considered to be a negative allosteric modulator of the GABA$_A$ receptor complex, and can enhance NMDA receptor activity via sigma receptors. Furthermore, DHEA can enhance memory formation in animal models. It is therefore possible that DHEA may play an important role in the enormous number of processes involved in memory formation and cognition.

Endogenous DHEA production is linked to age. There has been a great deal of interest in the role of DHEA in ageing, due to the effects on memory in animals. There are studies showing exciting effects of DHEA *in vitro* and in animal models for the treatment of learning and memory. These effects may be due to the conversion of DHEA to testosterone and estrogen, two very active steroid hormones. However, there is very little evidence from clinical studies to back-up claims that DHEA is the answer to the ageing process. Considering the new data showing that the human brain itself is capable of making DHEA, particularly in response to oxidative stress, further studies are required to determine the benefits, as well as the potential risks of DHEA supplementation in health and disease.

ACKNOWLEDGMENTS

This work was supported by a grant from the National Science Foundation (IBN 9728261). R.C.B. was supported by a National Science Foundation predoctoral fellowship. The authors would like to thank Drs. C. Cascio (University of Palermo, Palermo, Italy) and Z. Han (Georgetown University, Washington, DC, USA) for their contributions, Dr S. Lieberman (St. Luke's-Roosevelt Hospital Center, New York, NY, USA) for his encouragement and helpful discussions, Dr M. Culty (Georgetown University, Washington, DC, USA) for critically reviewing the manuscript, CIBA-Geigy (Basel, Switzerland) for the gift of SU 10603 and Stegram Pharmaceuticals (Sussex, UK) for the gift of Trilostane.

REFERENCES

Akwa, Y., Young, J., Kabbadj, K., Sancho, M.J., Zucman, D., Vourc'h, C. *et al.* (1991) Neurosteroids: biosynthesis, metabolism and function of pregnenolone and dehydroepiandrosterone in the brain. *J. Steroid Biochem. Molec. Biol.*, **40**, 71–81.

Albertson, B.D., Frederick, K.L., Maronian, N.C., Feuillan, P., Schorer, S., Dunn, J.F. *et al.* (1988a) The effect of ketoconazole on steroidogenesis: I. Leydig cell enzyme activity *in vitro*. *Res. Commun. Chem. Pathol. Pharmacol.*, **61**, 17–26.

Albertson, B.D., Maronian, N.C., Frederick, K.L., DiMattina, M., Feuillan, P., Dunn, J.F. *et al.* (1988b) The effect of ketoconazole on steroidogenesis: II. Adrenocortical enzyme activity *in vitro*. *Res. Commun. Chem. Pathol. Pharmacol.*, **61**, 27–34.

Arlt, W., Callies, F., van Vlijmen, J.C., Koehler, I., Reincke, M., Bidlingmaier, M. *et al.* (1999) Dehydroepiandrosterone replacement in women with adrenal insufficiency. *N. Engl. J. Med.*, **341**, 1013–1020.

Baulieu, E.E. (1996) Dehydroepiandrosterone (DHEA): a fountain of youth? *J. Clin. Endo. Metab.*, **81**, 3147–3151.

Baulieu, E.E. and Robel, P. (1998) Dehydroepiandrosterone (DHEA) and dehydro-epiandrosterone sulfate (DHEAS) as neuroactive neurosteroids. *Proc. Natl. Acad. Sci. USA*, **95**, 4089–4091.

Baulieu, E.E., Thomas, G., Legrain, S., Lahlou, N., Roger, M., Debuire, B. *et al.* (2000) Dehydroepiandrosterone (DHEA), DHEA sulfate, and aging: contribution of the DHEAge study to a sociobiomedical issue. *Proc. Natl. Acad. Sci. USA*, **11**, 4279–4284.

Beato, M. (1989) Gene regulation by steroid hormones. *Cell*, **56**, 335–344.

Bergeron, R., Montigny, C.D. and Debonnel, G. (1996) Potentiation of neuronal NMDA response induced by dehydroepiandrosterone and its suppression by progesterone: effects mediated by sigma receptors. *J. Neurosci.*, **16**, 1193–1202.

Berr, C., Lafont, S., Debuire, B., Dartigues, J.F. and Baulieu, E.E. (1996) Relationships of dehydroepiandrosterone sulfate in the elderly with functional, psychological, and mental status, and short-term mortality: a French community-based study. *Proc. Natl. Acad. Sci. USA*, **93**, 13410–13415.

Brown, R.C., Cascio, C. and Papadopoulos, V. (2000a) Pathways of neurosteroid biosynthesis in cell lines from human brain: regulation of dehydroepiandrosterone formation by oxidative stress and β-amyloid peptide. *J. Neurochem.*, **74**, 847–859.

Brown, R.C., Cascio, C. and Papadopoulos, V. (2000b) Neurosteroids: Oxidative stress mediated dehydroepiandrosterone formation in Alzheimer's disease. *Neurobiol. Aging*, **21**, S238.

Brown, R.C. and Papadopoulos, V. (2001) The role of the peripheral-type benzodiazepine receptor in adrenal and brain steroidogenesis. In R.H. Purdy and G. Biggio (eds), *Int. Rev. Neurobiol.*, **46**, 117–143.

Canning, D.R., McKeon, R.J., De Witt, D.A., Perry, G., Wujek, J.R., Frederickson, R.C. *et al.* (1993) β-Amyloid of Alzheimer's disease induces reactive gliosis that inhibits axonal outgrowth. *Exp. Neurol.*, **124**, 289–298.

Cardounel, A., Regelson, W. and Kalimi, M. (2000) Dehydroepiandrosterone protects hippocampal neurons against neurotoxin-induced cell death: mechanism of action. *Proc. Soc. Exp. Biol. Med.*, **222**, 145–149.

Cascio, C., Brown, R.C., Hales, D.B. and Papadopoulos, V. (2000) Pathways of dehydro-epiandrosterone formation in rat brain glia. *J. Steroid Biochem. Mol. Biol.*, **75**, 177–186.

Cascio, C., Guarneri, P., Li, H., Brown, R.C., Amri, H., Boujrad, N. *et al.* (1999) Peripheral-type benzodiazepine receptor. Role in the regulation of steroid and neurosteroid biosynthesis. In E.E. Baulieu, P. Robel and M. Schumacher (eds), *Neurosteroids: a new regulatory function in the central nervous system, contemporary endocrinology*, The Humana Press Inc., USA, pp. 75–96.

Cascio, C., Prasad, V.V.K., Lin, Y.Y., Lieberman, S. and Papadopoulos, V. (1998) Detection of P450c17-independent pathways for dehydroepiandrosterone (DHEA) biosynthesis in brain glial tumor cells. *Proc. Natl. Acad. Sci. USA*, **95**, 2862–2867.

Compagnone, N.A., Bulfone, A., Rubenstein, J.L.R. and Mellon, S.H. (1995) Steroidogenic enzyme P450c17 is expressed in the embryonic central nervous system. *Endocrinology*, **136**, 5212–5223.

Compagnone, N.A. and Mellon, S.H. (1998) Dehydroepiandrosterone: a potential signaling molecule for neocortical organization during development. *Proc. Natl. Acad. Sci. USA*, **95**, 4678–4683.

Corpechot, C., Robel, P., Axelson, M., Sjovall, J. and Baulieu, E.E. (1981) Characterization and measurement of dehydroepiandrosterone sulfate in rat brain. *Proc. Natl. Acad. Sci. USA*, **78**, 4704–4707.

Corpechot, C., Synguelakis, M., Talha, S., Axelson, M., Sjovall, J., Vihko, R. *et al.* (1983) Pregnenolone and its sulfate ester in the rat brain. *Brain Res.*, **270**, 119–125.

Corpechot, C., Young, J., Calvel, M., Wehrey, C., Veltz, J.N., Touyer, G. et al. (1993) Neurosteroids: 3α-hydroxy-5α-hregnan-20-one and its precursors in the brain, plasma, and steroidogenic glands of male and female rats. Endocrinology, 133, 1003–1009.

Cruess, D.G., Antoni, M.H., Kumar, M., Ironson, G., McCabe, P., Fernandez, J.B. et al. (1999) Cognitive-behavioral stress management buffers decreases in dehydroepiandrosterone sulfate (DHEA-S) and increases in the cortisol/DHEA-S ratio and reduces mood disturbance and perceived stress among HIV-seropositive men. Psychoneuroendocrinology, 24, 537–549.

Flood, J.F., Morley, J.E. and Roberts, E. (1992) Memory-enhancing effects in male mice of pregnenolone and steroids metabolically derived from it. Proc. Natl. Acad. Sci. USA, 89, 1567–1571.

Guazzo, E.P., Kirkpatrick, P.J., Goodyer, I.M., Shiers, H.M. and Herbert, J. (1996) Cortisol, dehydroepiandrosterone (DHEA) and DHEA sulfate in the cerebrospinal fluid of man: relation to blood levels and the effects of age. J. Clin. Endo. Metab., 81, 3951–3959.

Guennoun, R., Fiddes, R.J., Gouezou, M., Lombes, M. and Baulieu, E.E. (1995) A key enzyme in the biosynthesis of neurosteroids, 3β-hydroxysteroid dehydrogenase/delta 5-delta 4-isomerase (3β-HSD), is expressed in rat brain. Mol. Brain Res., 30, 287–300.

Hall, P.F. (1994) Cellular organization of steroidogenesis. Int. Rev. Cytol., 86, 53–95.

Hall, P.F. (1985) Role of cytochrome P-450 in the biosynthesis of steroid hormones. Vitamin. Horm., 42, 315–368.

Hu, Z.Y., Bourreau, E., Jung-Testas, I., Robel, P. and Baulieu, E.E. (1987) Neurosteroids: oligodendrocyte mitochondria convert cholesterol to pregnenolone. Proc. Natl. Acad. Sci. USA, 84, 8215–8219.

Hu, Z.Y., Jung-Testas, I., Robel, P. and Baulieu, E.E. (1989) Neurosteroids: steroidogenesis in primary cultures of rat glial cells after release of aminoglutethimide blockade. Biochem. Biophys. Res. Comm., 161, 917–922.

Huang, X., Atwood, C., Hartshorn, M., Multhaup, G., Goldstein, L., Scarpa, R. et al. (1999) The Aβ peptide of Alzheimer's disease directly produces hydrogen peroxide through metal ion reduction. Biochemistry, 38, 7609–7616.

Jung-Testas, I., Hu, Z.Y., Baulieu, E.E. and Robel, P. (1989) Neurosteroids: biosynthesis of pregnenolone and progesterone in primary cultures of rat glial cells. Endocrinology, 125, 2083–2091.

Kimonides, V.G., Khatibi, N.H., Svendsen, C.N., Sofroniew, M.V. and Hervert, J. (1998) Dehydroepiandrosterone (DHEA) and DHEA-sulfate (DHEA-S) protect hippocampal neurons against excitatory amino acid-induced neurotoxicity. Proc. Natl. Acad. Sci. USA, 95, 1852–1857.

Kohchi, C., Ukena, K. and Tsutsui, K. (1998) Age- and region-specific expressions of the messenger RNAs encoding for steroidogenic enzymes P450scc, P450c17 and 3β-HSD in the postnatal rat brain. Brain Res., 801, 233–238.

Kroboth, P., Salek, F., Pittenger, A., Fabian, T. and Frye, R. (1999) DHEA and DHEA-S: A review. J. Clin. Pharm., 39, 327–348.

LaCagnin, L., Levitt, M., Bergstrom, J. and Colby, H. (1989) Inhibition of adrenocortical, mitochondrial and microsomal monooxygenases by SU-10'603, a steroid 17α-hydroxylase inhibitor. J. Steroid Biochem., 33, 599–604.

Lambert, J.J., Belelli, D., Hill-Venning, C. and Peters, J.A. (1995) Neurosteroids and GABA_A receptor function. Trends Pharm. Sci., 16, 295–303.

LeGoascogne, C., Robel, P., Gouezou, M., Sananes, N., Baulieu, E.E. and Waterman, M. (1987) Neurosteroids: cytochrome P-450scc in rat brain. Science, 237, 1212–1215.

LeGoascogne, C., Sananes, N., Gouezou, M., Takemori, S., Kominami, S., Baulieu, E.E. et al. (1991) Immunoreactive cytochrome P-450(17 alpha) in rat and guinea pig gonads, adrenal glands and brain. J. Reprod. Fertil., 93, 609–622.

Lyne, C., Gower, D.B. and Lessof, M. (1974) Proceedings: effects of aminoglutethimide and SU 10'603 on the biosynthesis of C19 steroids. J. Endocrinol., 61, XVI–XVII.

Majewska, M.D. (1987) Steroids and brain activity. Essential dialogue between body and mind. *Biochem. Pharmacol.*, **36**, 3781–3788.

Majewska, M.D., Harrison, N.L., Schwartz, R.D., Barker, J.L. and Paul, S.M. (1986) Steroid hormone metabolites are barbiturate-like modulators of the GABA receptor. *Science*, **232**, 1004–1007.

Markesbery, W.R. (1997) Oxidative stress hypothesis in Alzheimer's disease. *Free Rad. Biol. Med.*, **23**, 134–147.

Mecocci, P., MacGarvey, U. and Beal, M. (1994) Oxidative damage to mitochondrial DNA is increased in Alzheimer's disease. *Ann. Neurol.*, **36**, 747–751.

Melcangi, R.C., Celotti, F. and Martini, L. (1994) Progesterone 5α reduction in neuronal and in different types of glial cell cultures: Type 1 and 2 astrocytes and oligodendrocytes. *Brain Res.*, **639**, 202–206.

Mellon, S.H. (1994) Neurosteroids: biochemistry, modes of action and clinical relevance. *J. Clin. Endo. Metab.*, **78**, 1003–1008.

Mellon, S.H. and Deschepper, C.F. (1993) Neurosteroid biosynthesis: genes for adrenal steroidogenic enzymes are expressed in the brain. *Brain Res.*, **629**, 283–292.

Mesiano, S., Coulter, C.L. and Jaffe, R.B. (1993) Localization of cytochrome P450 cholesterol side-chain cleavage, cytochrome P45017α-hydroxylase/17,20-lyase, and 3β-hydroxy-steroid Dehydrogenase isomerase steroidogenic enzymes in human and rhesus monkey fetal adrenal glands: reappraisal of functional zonation. *J. Clin. Endo. Metab.*, **77**, 1184–1189.

Miesfield, R.L. (1989) The structure and function of steroid receptor proteins. *Crit. Rev. Biochem. Mol. Biol.*, **24**, 101–117.

Miller, W.L., Auchus, R.J. and Geller, D.H. (1997) The regulation of 17,20 lyase activity. *Steroids*, **62**, 133–142.

Murray, M. and Reidy, G.F. (1990) Selectivity in the inhibition of mammalian cytochromes P-450 by chemical agents. *Pharmacol. Rev.*, **42**, 85–101

Nunomura, A., Perry, G., Pappolla, M., Wade, R., Hirai, K., Chiba, S. *et al.* (1999) RNA oxidation is a prominent feature of vulnerable neurons in Alzheimer's disease. *J. Neurosci.*, **19**, 1959–1964.

Orchinik, M. and McEwan, B. (1993) Novel and classical actions of neuroactive steroids. *Neurotransmissions*, **IX**, 1–6.

Papadopoulos, V., Guarneri, P., Krueger, K.E., Guidotti, A. and Costa, E. (1992) Pregnenolone biosynthesis in C6 glioma cell mitochondria: regulation by a diazepam binding inhibitor mitochondrial receptor. *Proc. Natl. Acad. Sc. USA*, **89**, 5113–5117.

Papadopoulos, V. (1993) Peripheral-type benzodiazepine/diazepam binding inhibitor receptor: biological role in steroidogenic cell function. *Endocr. Rev.*, **14**, 222–240.

Papadopoulos, V., Amri, H., Boujrad, N., Cascio, C., Culty, M., Garnier, M. *et al.* (1997) Peripheral benzodiazepine receptor in cholesterol transport and steroidogenesis. *Steroids*, **62**, 21–28.

Parker, C.J. (1999) Dehydroepiandrosterone and dehydroepiandrosterone sulfate production in the human adrenal during development and aging. *Steroids*, **64**, 640–647.

Prasad, V.V.K., Vegesna, S.R., Welch, M. and Lieberman, S. (1994) Precursors of the neuro-steroids. *Proc. Natl. Acad. Sci. USA*, **91**, 3220–3223.

Rainey, W.E., Bird, I.M., Sawetawan, C., Hanley, N.A., McCarthy, J.L., McGee, E.A. *et al.* (1993) Regulation of human adrenal carcinoma cell (NCI-H295) production of C19 steroids. *J. Clin. Endocrinol. Metab.*, **77**, 731–737.

Reiter, W.J., Pycha, A., Schatzl, G., Pokorny, A., Gruber, D.M., Huber, J.C. *et al.* (1999) Dehydroepiandrosterone in the treatment of erectile dysfunction: a prospective, double-blind, randomized, placebo-controlled study. *Urology*, **53**, 590–594.

Reiter, W.J., Pycha, A., Schatzl, G., Klingler, H.C., Mark, I., Auterith, A. *et al.* (2000) Serum dehydroepiandrosterone sulfate concentrations in men with erectile dysfunction. *Urology*, **55**, 755–758.

Robel, P., Young, J., Corpechot, C., Mayo, W., Perche, F., Haug, M. *et al.* (1995) Biosynthesis and assay of neurosteroids in rats and mice: functional correlates. *J. Steroid. Biochem. Molec. Biol.*, **53**, 355–360.

Roberts, E., Bologa, L., Flood, J.F. and Smith, G.E. (1987) Effects of dehydroepiandrosterone and its sulfate on brain tissue in culture and on memory in mice. *Brain Res.*, **406**, 357–362.

Romeo, E., Strohle, A., Spalletta, G., di Michele, F., Hermann, B., Holsboer, F. *et al.* (1998) Effects of antidepressant treatment on neuroactive steroids in major depression. *Am. J. Psychiatry*, **155**, 910–913.

Sanne, J.L. and Krueger, K. (1995) Aberrant splicing of rat steroid 17α-hydroxylase transcripts. *Gene*, **165**, 327–328.

Selye, H. (1941) Anesthetic effects of steroid hormones. *Proc. Soc. Exp. Biol.*, **46**, 116–121.

Selye, H. (1942) Correlations between the chemical structure and the pharmacological actions of the steroids. *Endocrinology*, **30**, 437–453.

Stocco, D.M. and Clark, B.J. (1996) Regulation of the acute production of steroids in steroidogenic cells. *Endocr. Rev.* **17**, 221–244.

Stromstedt, M. and Waterman, M.R. (1995) Messenger RNAs encoding steroidogenic enzymes are expressed in rodent brain. *Mol. Brain Res.*, **34**, 75–88.

Subbarao, K., Richardson, J. and Ang, L. (1990) Autopsy samples of Alzheimer's cortex show increased peroxidation *in vitro*. *J. Neurochem.*, **55**, 205–228.

Takebayashi, M., Kagaya, A., Uchitomi, Y., Kugaya, A., Muraoka, M., Yokota, N. *et al.* (1998) Plasma dehydroepiandrosterone sulfate in unipolar major depression. Short communication. *J. Neural Transm.*, **105**, 537–542.

Whitlock, J.P. (1986) The regulation of cytochrome P-450 gene expression. *Annu. Rev. Pharmacol. Toxicol.*, **26**, 333–369.

Wolf, O.T., Koster, B., Kirschbaum, C., Pietrowsky, R., Kern, W., Hellhammer, D.H. *et al.* (1997a) A single administration of dehydroepiandrosterone does not enhance memory performance in young healthy adults, but immediately reduces cortisol levels. *Biol. Psychiatry*, **42**, 845–848.

Wolf, O.T., Neumann, O., Hellhammer, D.H., Geiben, A.C., Strasburger, C.J., Dressendorger, R.A. *et al.* (1997b) Effects of a two-week physiological dehydroepiandrosterone substitution on cognitive performance and well being in healthy elderly women and men. *J. Clin. Endo. Metab.*, **82**, 2363–2367.

Wolkowitz, O.M., Reus, V.I., Keebler, A., Nelson, N., Friedland, M., Brizendine, L. *et al.* (1999) Double-blind treatment of major depression with dehydroepiandrosterone. *Am. J. Psychiatry*, **156**, 646–649.

Yoo, A., Harris, J. and Dubrovsky, B. (1996) Dose-response study of dehydroepiandrosterone sulfate on dentate gyrus long term potentiation. *Exp. Neurol.*, **137**, 151–156.

Zwain, I. and Yen, S. (1999) Dehydroepiandrosterone: biosynthesis and metabolism in the brain. *Endocrinology*, **140**, 880–887.

Chapter 9

Cross-talk between DHEA and neuropeptides

Chandan Prasad

INTRODUCTION

Dehydroepiandrosterone (DHEA) and DHEA.SO_4 (DHEAS) are two major steroid hormones produced and secreted by the adrenal gland of human and non-human primates (Miller, 1998; Kroboth *et al.*, 1999; Parker, 1999; Roberts, 1999). In contrast, the levels of these two steroid hormones in most other mammalian species are very low (Belanger *et al.*, 1989; Jo *et al.*, 1989; Guarneri *et al.*, 1994; Shimada *et al.*, 2000). Although the levels of DHEA and DHEAS are very low in the neuronal tissues, there are data to support that their formation or accumulation (or both) in the rat brain depends on *in situ* mechanisms unrelated to the peripheral endocrine gland (e.g. adrenal, placenta or ovary) (Corpechot *et al.*, 1981). Also there are data to support the contention that the glial cells of the brain are endowed with enzymatic machinery capable of converting cholesterol into DHEA and DHEAS (Corpechot *et al.* 1981; Shimada *et al.*, 2000). Both of these steroid hormones are not only endogenous to the brain, the administration of exogenous hormones to animals have been shown to exhibit a variety of central nervous system related biologic activities including appetite, memory, learning, mood, anxiety and depression to name a few (Svec *et al.*, 1998). Therefore, DHEA and DHEAS have been associated with a family of steroid hormones collectively known as "Neurosteroids".

This review will focus only on the relationship between neuropeptides and DHEA/DHEAS. A detailed critical review of the literature on synthesis, metabolism, and neurobiology of DHEA and DHEAS is presented elsewhere in this book. The major goal of this chapter is to review data on the nature of interdependence between DHEA and various neuropeptides. In brief, the review will include (i) how peptides of both neuronal and non-neuronal origin modulate synthesis and secretion of DHEA and DHEAS; (ii) how DHEA and DHEAS can affect synthesis and secretion of a variety of peptides known to modulate appetitive behavior; and lastly (iii) a hypothetical cross-link between DHEA/DHEAS and peptides.

PEPTIDES MODULATING DHEA/DHEAS SYNTHESIS

Pro-opiomelanocortin (POMC) peptides

The pro-opiomelanocortin gene codes for a precursor polypeptide (POMC) of molecular weight of approximately 31,000. POMC has been shown to be present in pituitary, brain, skin, and other peripheral sites of many species (Hadley and Haskell-Luevano, 1999). POMC is a precursor polypeptide that contains the amino acid sequences of numerous small polypeptide products (Bertagna, 1994). It contains eight pairs of basic amino acids and one sequence of four basic amino acids, which are the sites of cleavage for the recently identified prohormone convertases (PC1 and PC2), which have different specificities. PC1 but not PC2 is found in corticotroph cells of the anterior pituitary and cleaves POMC to generate *N*-terminal fragment (NT), joining peptide (JP), adrenocorticotropin (ACTH), β-lipotropin (β-LPH) and small amounts of γ-LPH and β-endorphin. In the melanotroph cells of the intermediate pituitary, both PC1 and PC2 are present and so therefore, proteolysis of POMC is more extensive, resulting in a series of smaller peptides including γ-melanocyte stimulating hormones (MSHs), corticotropin-like intermediate lobe peptide (CLIP or ACTH18–39), α-MSH, β-MSH, β-endorphin, and γ-endorphin. Equimolar amounts (because they are all products of the same precursor) of the POMC peptides are released into the blood by exocytosis from corticotroph cells of the anterior pituitary. No physiological role has been definitively identified for many of these products (Mountjoy and Wong, 1997).

Many POMC-derived peptides (e.g. JP, α-endorphin, β-endorphin, γ-endorphin, ACTH), and β-LPH and pro-enkephaline-derived peptides (Met-enkephaline and Leu-enkephaline) have been shown to regulate DHEA production by adrenal cells in culture (Hung and LeMaire, 1988; O'Connell *et al.*, 1993; Clarke *et al.*, 1996; Orso *et al.*, 1996; Mckenna *et al.*, 1997). The reticularis and fasciculata zones of the adrenal cortex are the predominant sources of DHEA and DHEA-SO$_4$ and contribute directly or indirectly 60–75% of androstenedione and testosterone in women. While the precise control of adrenal androgen remains unclear, β-endorphin and JP have been shown to stimulate androgen production in freshly prepared human adrenal cells and to influence ACTH-stimulated steroidogenesis in a manner that promotes adrenal androgen production (Clarke *et al.*, 1996; Mckenna *et al.*, 1997). JP was also found to enhance ACTH-stimulated DHEA production by human adrenocortical cells *in vitro* (Orso *et al.*, 1996). In freshly prepared guinea-pig adrenal cells *in vitro*, however, ACTH and β-LPH stimulated DHEA production whereas JP and α-, β- or γ-endorphins had no discernible effect (O'Connell *et al.*, 1993). In contrast, both Met-enkephaline and Leu-enkephaline inhibited DHEA production (O'Connell *et al.*, 1993). These results were not replicated when the effect of POMC-derived peptides was examined using monolayer tissue culture of mature female rat adrenal cells (Hung and LeMaire, 1988). In monolayer tissue culture of matured female rat adrenal cells, while ACTH was a potent stimulant for DHEA production, other peptides (Met-enkephaline, Leu-enkephaline, β-endorphin, β-LPH) had no measurable effect (Hung and LeMaire, 1988). Although the reasons for this anomaly are not apparent, it is conceivable that the adrenal cells may lose its ability to recognize these peptides due to culture conditions.

Leptin

Leptin is a 16-kDa adipocyte-secreted protein, the serum levels of which reflect mainly the amount of energy stores but are also influenced by short-term energy imbalance as well as several cytokines and hormones (Ahima *et al.*, 2000). Binding of leptin to its specific receptors, alters the expression of several hypothalamic neuropeptides that regulate neuroendocrine function as well as energy intake and expenditure (Campfield, 2000). More specifically, accumulating evidence suggests that this hormone may serve to signal to the brain information on the critical amount of fat stores that are necessary for lutenizing hormone releasing hormone (LHRH) secretion and activation of the hypothalamic-pituitary-gonadal axis (Mantzoros, 2000). Rising leptin levels have been associated with initiation of puberty in animals and humans and normal leptin levels are needed for maintenance of menstrual cycles and normal reproductive function (Mantzoros, 2000). Moreover, circadian and ultradian variations of leptin levels are associated with minute to minute variations of LH and estradiol in normal women (Mantzoros, 2000). Falling leptin levels in response to starvation result in decreased estradiol levels and amenorrhea in subjects with anorexia nervosa or strenuously exercising athletes (Mantzoros, 2000). In addition, leptin has been suggested to have a direct influence on ovarian steroidogenesis (Mantzoros, 2000).

In human adrenal cell preparations, leptin activates CYP17 enzyme activities (Biason-Lauber *et al.*, 2000). CYP17 is a microsomal enzyme consisting of two distinct activities – 17α-hydroxylase and 17,20-lyase – essential for the synthesis of cortisol and sex hormone precursors, respectively (Biason-Lauber *et al.*, 2000). The activation of CYP17, therefore, results in increased DHEA production. While leptin promotes DHEA synthesis (Biason-Lauber *et al.*, 2000), DHEAS decreases leptin secretion by female omental adipose tissue in humans (Pineiro *et al.*, 1999). This is also consistent with the observation that DHEA treatment decreases peripheral circulating leptin in obese Zucker characterized by hyperleptinemia and leptin resistance (Richards *et al.*, 2000). These data suggest that leptin and DHEA/DHEAS may function as counter-regulatory hormones.

DHEA/DHEAS AND APPETITE MODULATING PEPTIDES

Control of food intake is a complex phenomenon which is regulated by a variety of neurotransmitters and neuropeptides (Blundell, 1992; Blundell and King, 1996; Meguid *et al.*, 1996; Smith, 1999; Woods *et al.*, 2000). Some of these peptides inhibit food intake (e.g. cyclo (His-Pro), enterostatin, leptin and many others) whereas others stimulate (e.g. neuropeptide Y and others) (Blundell, 1992; Blundell and King, 1996; Meguid *et al.*, 1996; Smith, 1999; Woods *et al.*, 2000). Inhibition of food intake in rats and mice is one of the many known biologic activities associated with exogenous DHEA/DHEAS (Svec *et al.*, 1998). Nonetheless, it is difficult to ascertain the role for endogenous DHEA/DHEAS in appetite regulation since rats and mice produce little to none DHEA/DHEAS, (Belanger *et al.*, 1989; Jo *et al.*, 1989; Guarneri *et al.*, 1994; Shimada *et al.*, 2000). The role of DHEA/DHEAS in human appetitive behavior is equally uncertain since there is no data in the literature. However, results of a relatively indirect study suggest a possible inhibitory role for endogenous DHEAS in caloric intake in human females; in this study plasma concentration of DHEAS was 35% higher in "small-eaters" compared to the "large-eaters" subjects (Clark *et al.*, 1995). There are very few studies examining the relationships between DHEA and the peptides known to decrease caloric intake. The data presented below summarize the relationship between DHEA and two peptides, cyclo (His-Pro) and enterostatin. The relationship between DHEA and leptin is discussed earlier.

Cyclo(His-Pro)

Cyclo(His-Pro) or CHP is a peptide endogenous to the brain and a variety of other tissues in many species (Prasad, 1989, 1995). A role for endogenous CHP in appetite regulation is well documented (Prasad, 1988). In a recent study, we examined the regional brain distribution of CHP in hyperphagic obese Zucker rats and their lean littermates (Prasad et al., 1995). The data showed a significant elevation in the levels of CHP in many brain regions, including hypothalamus of the obese rat. Within the hypothalamus, the lateral hypothalamic (LH) nucleus of obese rats had significantly higher levels of CHP when compared to that of the lean littermates. Furthermore, administration of DHEA, a steroid hormone known to decrease food intake and body weight gain, to obese rats decreased the levels of CHP in the LH. This observation, however, is opposite to what empirical thinking would dictate. Both CHP and DHEA decrease food intake; therefore, one would expect DHEA to either raise CHP or have no effect.

Enterostatin

Enterostatins (Val-Pro-Asp-Pro-Arg (VPDPR), Val-Pro-Gly-Pro-Arg (VPGPR), and Ala-Pro-Gly-Pro-Arg (APGPR)) are pentapeptides derived from the NH2-terminus of procolipase after tryptic cleavage and belongs to the family of gut-brain peptides (Erlanson-Albertsson, 1992, 1994). Although enterostatin-like immunoreactivities exist in blood, brain, and gut (Sorhede et al., 1996a,b; Debata et al., 1998; Imamura et al., 1998a,b, 1999a,b), and exogenous enterostatins decrease fat appetite and insulin secretion in rats (Erlanson-Albertsson, 1992, 1994), the roles of these peptides in human obesity remain to be fully understood. In a recent study (Prasad et al., 1999) to understand whether VPDPR and VPGPR secretion is altered in obesity, serum VPDPR and VPGPR levels were measured in 38 overnight-fasted subjects (body mass index, 17.9–54.7 kg/m^2) before and after a meal. The mean fasting VPDPR in the serum of lean subjects was significantly lower than that in obese subjects (lean = 603±86 nmol/L (n = 17); obese, 1516±227 nmol/L (n = 21); $P = 0.0023$). In addition, the rise in serum VPGPR after a meal (post-meal/fasting ratio) was significantly higher in lean than in obese subjects (lean, 1.71±0.24 (n = 17); obese, 1.05±0.14 (n = 21); $P = 0.0332$). The results of these studies suggested hyperenterostatinemia in obesity and a diminution in enterostatin secretion after satiety.

DHEA is known to be a potent appetite suppressant in rodents (Svec et al., 1998). To understand the mechanism(s) of appetite modulation by DHEA, we have undertaken a series of studies to examine the effects of DHEA on neurotransmitters and neuropeptides known to affect appetitive behavior. In a recent study (Prasad et al., 2000), we examined the effect of DHEA on serum enterostatin-VPDPR or E, a pentapeptide known to cause selective diminution in fat intake.

Four-week-old lean (fa/+) and obese (fa/fa) Zucker rats were divided into control and treatment groups. DHEA-treated groups received powdered chow containing 0.6% DHEA *ad lib* for 16 weeks. Another group of obese rats was pair fed to match the intake of the obese DHEA-treated rats. At the end of this period, trunk blood was collected from fasted rats for assay of E-like immunoreactivity (E-LI) by ELISA. DHEA treatment caused a significant diminution in circulating E-LI in both lean (control: 2030 ± 226; treated: 752 ± 145 ng/mL; $n = 10$, $P < 0.0001$) and obese (control: 2489 ± 391, $n = 6$; treated: 1123 ± 185 ng/mL, $n = 7$; $P = 0.0003$) rats. Because DHEA treatment decreases caloric intake and body weight, we examined the effect of caloric intake and body weight on E-LI levels. Serum E-LI levels were lower in the obese DHEA-treated group compared to that of the obese pair fed (pair fed: 1589 ± 313, $n = 6$; DHEA: 1123 ± 185 ng/mL, $n = 7$), but the differences were statistically insignificant ($P = 0.185$). Also, both weight-matched lean and obese control rats had significantly ($P < 0.008$) higher E-LI than their DHEA-treated counterparts. To examine whether the decrease in serum E-LI following DHEA treatment could be due to increased peptide metabolism, the rate of disappearance of endogenous E-LI from serum (obese control and DHEA-treated) at $37°C$ was evaluated. The results show an attenuation of peptide metabolism in serum from DHEA-treated rats, a finding contrary to our expectations. In summary, DHEA treatment lowers serum E-LI levels both in lean and obese Zucker rats. This decrement in peptide level is not secondary to changes in body weight or caloric intake due to DHEA, or due to altered serum peptide metabolism. Although DHEA appears to be a potent modulator of E-LI levels, the relationship between DHEA and E-LI in relation to appetitive behavior remains to be clarified.

DHEA/DHEAS, PEPTIDES, AND CALORIC INTAKE – SEARCH FOR A LINK

Over the past several years, it has become evident that in addition to classical neurotransmitters – dopamine, norepinephrine, serotonin and others – peptides play an important role in regulation of caloric intake. A detailed survey of the literature on the relationship between neurotransmitters and DHEA in relation to appetite control is presented elsewhere in this book. At present there are over a dozen peptides known to decrease or increase food intake. By simple logic, one would assume that the levels of the peptides known to decrease appetite would be lower in an obese state which is often associated with increased caloric intake. In contrast, however, the level of 4 peptides (see Table 9.1) that we have looked at were elevated in obesity. It is tempting to hypothesize that elevation in the levels of appetite inhibiting peptides in obesity is secondary to the development of a resistant state i.e. these peptides do not function effectively in the obese state. Consequently, to counteract the resistant state, the system produces more and more of these peptides. Thus, increase in the level of these peptides is not the cause of obesity, but the consequence of obesity. However, the relationship between DHEA, an appetite suppressant, and these peptides also appears far from simple. The levels of cyclo (His-Pro), enterostatin, leptin, and insulin, peptides/polypeptides known to decrease caloric intake, are elevated in the obese state and treatment with

Table 9.1 Relationship between DHEA and peptides inhibiting caloric intake

Peptides	Effect on food intake	Level in obesity	Effect of DHEA
Cyclo(His-Pro)	↓	↑	↓
Enterostatin	↓	↑	↓
Leptin	↓	↑	↓
Insulin	↓	↑	↓

Note
↑ increased; ↓ decreased

DHEA, an agent known to decrease caloric intake, decreases the levels of all of these peptides. If the relationships were simple, once again, DHEA would have been expected to raise the levels of these peptides. Again, a hypothetical explanation for these data will be: DHEA lowers caloric intake/obesity by a mechanism independent of changes in the levels of these peptides; decrease in obesity lowers peptide level and increases peptide sensitivity (or decreases peptide resistance), which in turn makes the endogenous peptide more effective. One way to test this hypothesis would be to show that DHEA treatment increases the efficacy of cyclo (His-Pro), enterostatin, leptin, or insulin in reducing caloric intake.

REFERENCES

Ahima, R.S., Saper, C.B., Flier, J.S. and Elmquist, J.K. (2000) Leptin regulation of neuro-endocrine systems. *Front. Neuroendocrin.*, **21**, 263–307.

Belanger, B., Belanger, A., Labrie, F., Dupont, A., Cusan, L. and Monfette, G. (1989) Comparison of residual C-19 steroids in plasma and prostatic tissue of human, rat and guinea pig after castration: unique importance of extratesticular androgens in men. *J. Steroid Biochem.*, **32**, 695–698.

Bertagna, X. (1994) Proopiomelanocortin-derived peptides. *Endocrin. Metab. Clin. N. Amer.*, **23**, 467–485.

Biason-Lauber, A., Zachmann, M. and Schoenle, E.J. (2000) Effect of leptin on CYP17 enzymatic activities in human adrenal cells: new insight in the onset of adrenarche. *Endocrinology*, **141**, 1446–1454.

Blundell, J.E. (1992) Serotonin and the biology of feeding. *Am. J. Clin. Nutr.*, **55**, 155S–159S.

Blundell, J.E. and King, N.A. (1996) Overconsumption as a cause of weight gain: behavioral–physiological interactions in the control of food intake (appetite). *Ciba F. Symp.*, **201**, 138–154; discussion 154–158, 188–193.

Campfield, L.A. (2000) Central mechanisms responsible for the actions of OB protein (leptin) on food intake, metabolism and body energy storage. *Front. Horm. Res.*, **26**, 12–20.

Clark, D.G., Tomas, F.M., Withers, R.T., Brinkman, M., Berry, M.N., Oliver, J.R. *et al.* (1995) Differences in substrate metabolism between self-perceived "large-eating" and "small-eating" women. *Int. J. Obesity Rel. Metab. Disord.*, **19**, 245–252.

Clarke, D., Fearon, U., Cunningham, S.K. and McKenna, T.J. (1996) The steroidogenic effects of beta-endorphin and joining peptide: a potential role in the modulation of adrenal androgen production. *J. Endocrinol.*, **151**, 301–307.

Corpechot, C., Robel, P., Axelson, M., Sjovall, J. and Baulieu, E.E. (1981) Characterization and measurement of dehydroepiandrosterone sulfate in rat brain. *Proc. Natl. Acad. Sci. USA*, **78**, 4704–4707.

Debata, C. and Prasad, C. (1998) Endogenous enterostatin, proteases, and dietary fat preference. *Nutr. Neurosci.*, **1**, 361–366.

Erlanson-Albertsson, C. (1992) Enterostatin: the pancreatic procolipase activation peptide – a signal for regulation of fat intake. *Nutr. Rev.*, **50**, 307–310.

Erlanson-Albertsson, C. (1994) Pancreatic lipase, colipase and enterostatin – a lipolytic triad. In M.I. Mackness and M. Clerc (eds), *Esterases, lipases and phospholipases: from structure to clinical significance*, Plenum Press, New York, pp. 159–168.

Guarneri, P., Guarneri, R., Cascio, C., Pavasant, P., Piccoli, F. and Papadopoulos, V. (1994) Neurosteroidogenesis in rat retinas. *J. Neurochem.*, **63**, 86–96.

Hadley, M.E. and Haskell-Luevano, C. (1999) The proopiomelanocortin system. *Ann. N. Y. Acad. Sci.*, **885**, 1–21.

Hung, T.T. and LeMaire, W.J. (1988) The effects of corticotropin, opioid peptides and crude pituitary extract on the production of dehydroepiandrosterone and corticosterone by mature rat adrenal cells in tissue culture. *J. Steroid Biochem.*, **29**, 721–726.

Imamura, M., Debata, C. and Prasad, C. (1999a) On the nature and distribution of enterostatin (Val-Asp-Pro-Asp-Arg)-like immunoreactivity in rat plasma. *Peptides*, **20**, 133–139.

Imamura, M., Porter, J.R. and Prasad, C. (1999b) Differential regional distribution of enterostatin, an appetite inhibiting peptide, in the brains of Zucker and Sprague-Dawley rats. *Nutr. Neurosci.*, **1**, 449–453.

Imamura, M., Porter, J.R. and Prasad, C. (1998a) Differential regional distribution of enterostatin, an appetite inhibiting peptide, in the brains of Zucker and Sprague-Dawley rats. *Nutr. Neurosci.*, **1**, 449–453.

Imamura, M., Sumar, N., Hermon-Taylor, J., Robertson, H.J.F. and Prasad, C. (1998b) Distribution and characterization of enterostatin-like immunoreactivity in human cerebrospinal fluid. *Peptides*, **19**, 1385–1391.

Jo, D.H., Abdallah, M.A., Young, J., Baulieu, E.E. and Robel, P. (1989) Pregnenolone, dehydroepiandrosterone, and their sulfate and fatty acid esters in the rat brain. *Steroids*, **54**, 287–297.

Kroboth, P.D., Salek, F.S., Pittenger, A.L., Fabian, T.J. and Frye, R.F. (1999) DHEA and DHEA-S: a review. *J. Clin. Pharmacol.*, **39**, 327–348.

Mantzoros, C.S. (2000) Role of leptin in reproduction. *Ann. N. Y. Acad. Sci.*, **900**, 174–183.

McKenna, T.J., Fearon, U., Clarke, D. and Cunningham, S.K. (1997) A critical review of the origin and control of adrenal androgens. *Bailliere Clin. Ob. Gy.*, **11**, 229–248.

Meguid, M.M., Yang, Z.J. and Gleason, J.R. (1996) The gut-brain brain-gut axis in anorexia: toward an understanding of food intake regulation. *Nutrition*, **12**, S57–S62.

Miller, W.L. (1998) Early steps in androgen biosynthesis: from cholesterol to DHEA. *Bailliere Clin. Endocr. Metab.*, **12**, 67–81.

Mountjoy, K.G. and Wong, J. (1997) Obesity, diabetes and functions for proopiomelanocortin-derived peptides. *Mol. Cell. Endocrinol.*, **128**, 171–177.

O'Connell, Y., McKenna, T.J. and Cunningham, S.K. (1993) Effects of pro-opiomelanocortin-derived peptides on adrenal steroidogenesis in guinea-pig adrenal cells *in vitro*. *J. Steroid Biochem. Mol. Biol.*, **44**, 77–83.

Orso, E., Szalay, K.S., Szabo, D., Stark, E., Feher, T., Perner, F. *et al.* (1996) Effects of joining peptide (1–18) and histamine on dehydroepiandrosterone (DHEA) and dehydroepiandrosterone sulphate (DHEAS) production of human adrenocortical cells *in vitro*. *J. Steroid Biochem. Mol. Biol.*, **58**, 207–210.

Parker, Jr. C.R. (1999) Dehydroepiandrosterone and dehydroepiandrosterone sulfate production in the human adrenal during development and aging. *Steroids*, **64**, 640–647.

Pineiro, V., Casabiell, X., Peino, R., Lage, M., Camina, J.P., Menendez, C. *et al.* (1999) Dihydrotestosterone, stanozolol, androstenedione and dehydroepiandrosterone sulphate inhibit leptin secretion in female but not in male samples of omental adipose tissue *in vitro*: lack of effect of testosterone. *J. Endocrinol.*, **160**, 425–432.

Prasad, A., Richards, R.J., Svec, F., Porter, J.R. and Prasad, C. (2000) Dehydroepiandro-sterone-mediated decrease in caloric intake by obese Zucker rats is not due to changes in serum entrostatin-like immunoreactivity. *Physiol. Behav.*, **68**, 341–345.

Prasad, C. (1988) Cyclo(His-Pro): its distribution, origin, and function in the human. *Neurosci. Biobehav. Rev.*, **12**, 19–22.

Prasad, C. (1989) Neurobiology of cyclo(His-Pro). *Ann. N. Y. Acad. Sci.*, **553**, 232–251.

Prasad, C. (1995) Bioactive cyclic dipeptides. *Peptides*, **16**, 151–164.

Prasad, C., Imamura, M., Debata, C., Svec, F., Sumar, N. and Hermon-Taylor, J. (1999) Hyperenterostatinemia in premenopausal obese women. *J. Clin. Endocr. Metab.*, **84**, 937–941.

Prasad, C., Mizuma, H., Brock, J.W., Porter, J.R., Svec, F. and Hilton, C. (1995) A paradoxical elevation of brain cyclo(His-Pro) levels in hyperphagic obese Zucker rats. *Brain Res.*, **699**, 149–153.

Richards, R.J., Porter, J.R. and Svec, F. (2000) Serum leptin, lipids, free fatty acids, and fat pads in long-term dehydroepiandrosterone-treated Zucker rats. *Proc. Soc. Exp. Biol. Med.*, **223**, 258–262.

Roberts, E. (1999) The importance of being dehydroepiandrosterone sulfate (in the blood of primates): a longer and healthier life? *Biochem. Pharmacol.*, **57**, 329–346.

Schwartz, M.W., Woods, S.C., Porte, Jr. D., Seeley, R.J. and Baskin, D.G. (2000) Central nervous system control of food intake. *Nature*, **404**, 661–671.

Shimada, K. and Yago, K. (2000) Studies on neurosteroids X. Determination of preg-nenolone and dehydroepiandrosterone in rat brains using gas chromatography-mass spectrometry-mass spectrometry. *J. Chromatogr. Sci.*, **38**, 6–10.

Smith, G.P. (1999) Introduction to the reviews on peptides and the control of food intake and body weight. *Neuropeptides*, **33**, 323–328.

Sorhede, M., Erlanson-Albertsson, C., Mei, J., Nevalainen, T., Aho, A. and Sundler, F. (1996a) Enterostatin in gut endocrine cells – immunocytochemical evidence. *Peptides*, **17**, 609–614.

Sorhede, M., Mulder, H., Mei, J., Sundler, F. and Erlanson-Albertsson, C. (1996b) Pro-colipase is produced in the rat stomach – a novel source of enterostatin. *Biochim. Biophys. Acta*, **1301**, 207–212.

Svec, F. and Porter, J.R. (1998) Dehydroepiandrosterone: a nutritional supplement with actions in the central nervous system. *Nutr. Neurosci.*, **1**, 9–19.

Svec, F., Richards, R.J. and Porter, J.R. (1998) Investigating the debate: does DHEA alter food intake? *Nutr. Neurosci.*, **1**, 93–101.

Woods, S.C., Schwartz, M.W., Baskin, D.G. and Seeley, R.J. (2000) Food intake and the regulation of body weight. *Annu. Rev. Psychol.*, **51**, 255–277.

Chapter 10

DHEA effects on human behavior and psychiatric illness

Victor I. Reus and Owen M. Wolkowitz

INTRODUCTION

Given the scope of recent interest in the neuropsychiatric effects of DHEA/S, it is somewhat surprising to learn that the first published reports appeared in the early 1950's (Sands and Chamberlain, 1952; Strauss *et al.*, 1952). For unclear reasons public and professional interest in human behavioral effects waned until the late 1980's, at which point a confluence of animal behavioral and human epidemiologic data suggested that DHEA may naturally regulate aspects of human mood, cognition, and behavior and possibly counteract many of the adverse effects of ageing. The positive results reported in the first well controlled clinical trial of the pharmacologic effects of DHEA replacement in humans (Morales *et al.*, 1994) further kindled scientific investigation and enthusiasm. Although recent studies have been somewhat more mixed in their conclusions, a wealth of evidence now exists that indicates that DHEA/S levels frequently correlate with mood, psychiatric status, memory and functional abilities in humans and that pharmacologic treatment with DHEA may have salutary effects on these same domains in normal individuals, as well as various patient groups. This chapter chronicles the progress that has been made in understanding the behavioral role of DHEA regulation and treatment in humans.

ASSOCIATION BETWEEN DHEA/S LEVEL AND MOOD, COGNITION, AND FUNCTIONAL ABILITIES IN NORMAL INDIVIDUALS

A variety of studies have investigated the relationship between DHEA/S levels and measures of health, global functioning, cognition and overall "well being" in both cross-sectional and longitudinal designs. Rudman *et al.* (1990) and Morrison *et al.* (1998) reported that lower levels of DHEA/S were found in more frail individuals drawn from nursing home population, a finding similar to that reported by Berkman *et al.* (1993), who studied high, median, and low functioning individuals in the community. Higher DHEA/S levels were observed in a high functioning subset of the "oldest-old," i.e. individuals over 90 years of age (Ravaglia *et al.*, 1996). In a large epidemiologic study conducted in south-west France, significantly lower levels of DHEA/S were recorded in women who had functional limitations, depressive symptoms and poorer self perceptions of health and life satisfaction (Berr *et al.*, 1996).

The possible relationship between DHEA/S and cognition seems more obscure. Barrett-Connor and Edelstein (1994), in a prospective study of healthy elderly community dwelling subjects, found that DHEA/S was not a prediction of cognitive change over time, although DHEA/S levels were inversely associated with depressed mood (Barrett-Connor *et al.*, 1999). Similar findings were reported by Yaffe *et al.* (1998), but Kalmijn *et al.* (1998) concluded that lower levels did indeed increase odds of becoming cognitively impaired. The opposite result has also been reported, i.e. that higher DHEA/S levels are associated with greater cognitive impairment (Morrison *et al.*, 2000).

Cawood and colleagues (Cawood and Bancroft, 1996) did not directly assess cognition, but observed that DHEA level (and not DHEAS) was significantly and positively correlated with ratings of "well being" and positive affect in a group of community based women aged 40–60. A higher cortisol to DHEAS ratio was observed in women experiencing greater depression and anxiety during menopause (Tode *et al.*, 1999). Studies of DHEA/S relationship to mood have also been conducted in younger groups. Buckwalter *et al.* (1999) for example, assayed neuropsychological functioning in 19 women over the course of their pregnancies and observed that higher levels of DHEA were associated with better mood.

BEHAVIORAL EFFECTS OF DHEA ADMINISTRATION IN NORMAL INDIVIDUALS

The correlations noted between DHEA/S, mood and functional status have led to experimental trials of acute and more extended pharmacologic treatment with the compound. Many of the first behavioral effects were noted serendipitously in the context of DHEA treatment of primary medical disorders. In studies of open administration of DHEA to patients with multiple sclerosis (Calabrese *et al.*, 1990), systemic lupus erythematosis (SLE) (van Vollenhoven *et al.*, 1994) and myotonic dystrophy (Sugino *et al.*, 1998), patients exhibited increased libido, energy and improved mood and daily functional ability. DHEA replacement in patients with adrenal insufficiency increased well being and sexuality, while DHEA treatment in HIV positive patients resulted in decreased fatigue and depression (Arlt *et al.*, 1998; Rabkin *et al.*, 2000). Cognitive-behavioral stress management buffered decreases in DHEAS and also reduced mood disturbance and perceived stress in HIV positive men (Cruess *et al.*, 1999).

More recently, Morales *et al.* (1994) gave 50 mg/day of DHEA or placebo for three months to a group of middle aged men and women in a randomized, cross over design. Individuals receiving active drug reported significant improvement in "well being", which was individually interpreted as secondary to beneficial changes in energy, sleep, relaxation, and stress tolerance. A specific increase in insulin-like growth factor-I (IGF-I) accompanied the behavioral change and was hypothesized to be causally related. Unfortunately, a longer term (6 month) treatment study, using a larger dose (100 mg/day), by the same group was unable to replicate the dramatic behavioral response initially reported (Morales *et al.*, 1998).

Other placebo controlled studies employing similar designs and populations have also reported negative findings, although it is unclear if the behavioral instruments used were sufficiently sensitive (Casson *et al.*, 1998; Flynn *et al.*, 1999). Acute administration of DHEA, even in high dose, was also not found to improve self rated mood, fatigue or cognition in a controlled study with young adults (Wolf *et al.*, 1997a). In an uncontrolled effort, however, trans-dermal application of DHEA for 12 months reportedly improved well being in menopausal women (Diamond *et al.*, 1996). In general, positive behavioral responses to DHEA appear more frequently in investigations utilizing older cohorts. Wolf and colleagues, in a series of reports (Wolf *et al.*, 1997b, 1998a,b) did not find consistent effects of drug on memory performance, but did note improvements in mood in women, enhanced P300 amplitude on evoked response testing, and improved attention response. However, the drug administration period was limited to two weeks, leaving open the question as to whether a longer treatment period may have produced a different profile. Effects of DHEA on the behavioral symptomatology of menopause are similarly inconclusive, with both positive (Stomati *et al.*, 1999) and negative reports in the literature (Barnhart *et al.*, 1999).

In summary, the behavioral effects of DHEA treatment in normal individuals remain obscure. Variations in study design, in dosage and duration of treatment, in populations studied and instruments used limit the conclusions that can be drawn (Huppert *et al.*, 2000). Clearly, longer term, large scale studies focused on an elderly population and utilizing sophisticated behavioral assessments are warranted. Two such investigations are currently underway (Legrain *et al.*, 2000; L. Brizendine, personal communication).

DHEA/S IN DEPRESSION

The extant literature examining levels of serum or urinary DHEA/S in depressed patients is an inconsistent one, with findings of decreased (Ferguson *et al.*, 1964; Heinz *et al.*, 1999; Michael *et al.*, 2000), increased (Hansen *et al.*, 1982; Tollefson *et al.*, 1990; Heuser *et al.*, 1998; Takebayashi *et al.*, 1998), and no change in value compared to control populations (Reus *et al.*, 1993; Fava *et al.*, 1989; Shulman *et al.*, 1992). The relationship between DHEA and cortisol may provide a better marker of mood state, given the Goodyer *et al.* (1996) report of DHEA hyposecretion and evening cortisol hypersecretion in adolescents with major depression and similar findings by Osran *et al.* (1993) and Ferrari *et al.* (1997) in adults. In adolescents with major depression who were followed for one year, low night-time ratios of DHEA to cortisol predicted persistence of depression and recurrent adverse life events. A recent paper by this same group, however, also found that early morning hypersecretion of DHEA was associated with the development of major depression in a high risk cohort followed over 12 months (Goodyer *et al.*, 2000). This would indicate that altered circadian patterning of release may need to be considered in interpretation of conflictual findings. Heuser's study (Heuser *et al.*, 1998) is particularly informative in that it was based on sampling every 30 minutes over a 24 hour period. Depressed patients had increased diurnal minimal and mean DHEA plasma concentrations, but no change in maximal levels.

Correlations have been found in the ratio between DHEA and cortisol and their synthetic enzymes and symptom severity in depression as well (Raven and Taylor, 1998). Eleven-beta-hydroxysteroid dehydrogenase (HSD) activity correlated with degree of depression in women, while 17-beta HSD showed a significant positive correlation in women and a negative association in men.

The effects of antidepressant treatment have also been evaluated. Here again the results are difficult to understand: Electroconvulsive treatment (ECT), for example, increased DHEAS levels (Ferguson *et al.*, 1964), while markedly elevated basal levels are associated with resistance to ECT (Maayan *et al.*, 2000). Tricyclic antidepressants, such as imipramine (Tollefson *et al.*, 1990) or clomipramine (Takebayashi *et al.*, 1998) lowered plasma and/or urinary DHEAS levels, but Romeo *et al.* (1998) reported no effect for antidepressant drugs on plasma DHEA concentration and Uzunova *et al.* (1998), no change in CSF DHEA/S in response to treatment with fluoxetine or paroxetine. Alprazolam, a triazolobenzodiazepine, increased plasma DHEA, but not DHEAS levels, and those subjects who showed the greatest increase also experienced lesser psychomotor impairment, suggesting that DHEA may be being converted to other neurosteroid compounds having $GABA_A$ receptor agonist effects (Kroboth *et al.*, 1999).

Psychosocial treatments in general have been reported to result in increases in DHEA/S. Interventions that have been evaluated include cognitive-behavioral treatment of depression (Tollefson *et al.*, 1990), stress reduction (Littman *et al.*, 1993), "social enrichment" (Arnetz *et al.*, 1983; McCarty *et al.*, 1998), and trancendental meditation (Glaser *et al.*, 1992).

Studies of beneficial mood improvement with DHEA in normal volunteers have recently resulted in controlled investigations of therapeutic efficacy in major depression. In an initial open trial, Wolkowitz *et al.* (1997) gave 30–90 mg of DHEA for four weeks to six middle aged and elderly patients with major depression. Objective and subjective measures of mood state showed significant improvement, as did a measure of "automatic" memory. One individual received extended treatment for six months and demonstrated continued improvement in depression ratings and in semantic memory performance. Most interestingly, improvements in depression ratings and memory were directly related to increases in plasma levels of DHEA and DHEAS and to their ratios with cortisol. In a follow-up investigation, Wolkowitz *et al.* (1999a) administered DHEA (up to 90 mg/day) or placebo in a random fashion to a cohort of patients with major depression for six weeks. Patients were either drug free or stabilized on their pre-existing medication at baseline. Five of 11 subjects showed a 50% or greater decrease in depressive symptoms, compared to none on placebo. A similar robust antidepressant effect was recently reported by Bloch and colleagues (Bloch *et al.*, 1999) in a study in which 15 patients with dysthymia were given 90 mg, and then 450 mg of DHEA, for three weeks each or placebo in a randomized cross-over design. Sixty percent of subjects responded, as compared to 20% on placebo, with significant effects noted after only three weeks on the 90 mg dosage. DHEA was well tolerated in both studies. DHEA treatment may, however, carry the risk of precipitating mania in susceptible individuals (Markowitz *et al.*, 1999; Kline and Jaggers, 1999). Taken together, these studies suggest that DHEA, either alone or administered as an adjunctive agent, may act as an effective antidepressant agent. Larger scale attempts at replication appear warranted.

DHEA IN OTHER PSYCHIATRIC AND STRESSFUL CONDITIONS

A variety of reports have looked at correlations between measures of temperament or symptomatic state and DHEA. Elevations in perceived stress (Labbate *et al.*, 1995), trait anxiety (Diamond *et al.*, 1989) and cynicism, hostility and Type A behavior (Fava *et al.*, 1987, 1992; Littman *et al.*, 1993) appear to go along with lower levels of DHEA/S, while high DHEA levels were related to greater anger and dominance in ageing males (Gray *et al.*, 1991), and pre-pubertal boys with conduct disorder (van Goozen *et al.*, 1998), they appeared to be related to less aggressive affect in 10–14 year old girls (Brooks-Gunn and Warren, 1989). Negative findings also exist, however (Constantino *et al.*, 1993).

Other isolated findings include reports of an increased DHEAS to cortisol ratio in patients with panic disorder (Fava *et al.*, 1989), and very low DHEA levels and DHEA to cortisol ratios in anorexia nervosa (Winterer *et al.*, 1985; Zumoff *et al.*, 1983). Although the ratios appear to normalize with weight gain, DHEA treatment had no obvious beneficial effects on either weight or mood (Gordon *et al.*, 1999).

In patients diagnosed with cocaine abuse, DHEA to cortisol ratios have been used to discriminate individual differences in vulnerability to relapse (Majewska *et al.*, 1997). In patients with alcohol dependence, levels of DHEAS and DHEAS/cortisol ratio were found to be inversely correlated with depression and perhaps involved with subsequent risk of relapse (Heinz *et al.*, 1999).

Regulation of DHEA/S has also been evaluated in patients with schizophrenia. Again, some investigations have reported elevated DHEAS levels (Oades and Schep-ker, 1994) and others decreased DHEA (Dilbaz *et al.*, 1998; Tourney and Hatfield, 1972). Erb *et al.* (1981), using a cut off level of <470 ng/dl for plasma DHEA (8:00 am), reported that they could discriminate 85% of their patients from controls. A similar observation was recently made by Harris and colleagues (Harris *et al.*, 2001). In this study, morning DHEA level and DHEA/cortisol ratio were positively correlated with memory performance and negatively correlated with measures of psychosis and extrapyramidal symptoms in a group of 17 chronic schizophrenic inpatients at a state hospital. The potential mechanism for such an association, if real, remains speculative, but may involve decreased $GABA_A$ receptor function.

DHEA IN DEMENTIA

Measurements of DHEA/S levels in patients with Alzheimer's disease are equivocal (Miller *et al.*, 1998). Some studies, such as Sunderland *et al.* (1989) have reported significantly lower levels (Nasman *et al.*, 1991; Yanase *et al.*, 1996; Solerte *et al.*, 1999; Hillen *et al.*, 2000; Murialdo *et al.*, 2000) while others find no specific pattern of change (Leblhuber *et al.*, 1993; Legrain *et al.*, 1995; Schneider and Hinsey, 1992; Spath-Schwalbe *et al.*, 1990). Patients with multi-infarct dementia have similarly low levels in several of these investigations. Although no overall differences in DHEAS or cortisol levels, or in DHEAS to cortisol ratio were found between Alzheimer's patients and normal controls by Carlson *et al.* (1999), those demented patients who had higher levels of DHEAS scored better on a battery of cognitive tests than did patients with lower values. The inconsistency in published findings has been attributed to possible variability in the rate of metabolism of DHEA to 7-alpha-hydroxy DHEA, a compound with antiglucocorticoid effects (Attal-Kemis *et al.*, 1998). Other investigators have looked to longitudinal designs as more relevant to the core issue than the more common cross-sectional comparisons. In this regard Lupien *et al.* (1999) found that greater decreases in DHEAS to cortisol ratio predicted more obvious cognitive deterioration in a group of elderly men and women. Others, however, have been unable to replicate this (Barret-Connor and Edelstein, 1994; Miller *et al.*, 1998; Yaffe *et al.*, 1998).

Even if dementia involves a specific change in DHEA/S regulation over that observed in normal ageing, the key question is whether pharmacologic replacement would have any therapeutic benefit. In a recent completed study, Wolkowitz *et al.* (1999b) gave DHEA (50 mg, orally, twice a day) or placebo to 58 subjects with Alzheimer's disease over six months. Active treatment, relative to placebo, was associated with significant improvement in cognitive function after three months and a trend towards improvement by month six. Significantly, more DHEA treated subjects than did placebo treated individuals. If confirmed, these data indicate that restoring DHEA and DHEAS levels to values seen in early adulthood may benefit patients with Alzheimer's dementia.

CONCLUSION

A critical reading of the literature must conclude that the role of DHEA/S in neuropsychiatric disease and its possible utility in alteration of behavior remain uncertain. Although a great number of studies have been completed, methodological differences in design, hormonal sampling, and outcome measures continue to limit the conclusions that can be drawn. In addition to questions of clinical efficacy, the long-term risks associated with DHEA/S administration remain unknown. Positive findings, deriving from well controlled investigations, continue to be of interest, however, and encourage the view that future studies, employing larger number of subjects and longer periods of treatment, will be illuminating.

REFERENCES

Arlt, W., Justl, H.G., Callies, F., Reincke, M., Hubler, D., Oettel, M. *et al.* (1998) Oral dehydro-epiandrosterone for adrenal androgen replacement: pharmacokinetics and peripheral conversion to androgens and estrogens in young healthy females after dexamethasone suppression. *J. Clin. Endocrinol. Metab.*, **83**, 1928–1934.

Arnetz, B.B., Theorell, T., Levi, L., Kallner, A. and Eneroth, P. (1983) An experimental study of social isolation of elderly people: psychoendocrine and metabolic effects. *Psychosomatic Med.*, **45**, 395–406.

Attal-Khemis, S., Dalmeyda, V. and Morfin, R. (1998) Change of 7alpha-hydroxy-dehydro-epiandrosterone levels in serum of mice treated by cytochrome P450-modifying agents. *Life Sci.*, **63**, 1543–1553.

Barnhart, K.T., Freeman, E., Grisso, J.A., Rader, D.J., Sammel, M., Kapoor, S. *et al.* (1999) The effect of dehydroepiandrosterone supplementation to symptomatic perimenopausal women on serum endocrine profiles, lipid parameters, and health-related quality of life. *J. Clin. Endocrinol. Metab.*, **84**, 3896–3902.

Barrett-Connor, E. and Edelstein, S.L. (1994) A prospective study of dehydroepiandrosterone sulfate and cognitive function in an older population: the Rancho Bernardo study. *J. Am. Geriatr. Soc.*, **42**, 420–423.

Barrett-Connor, E., Von Muhlen, D.G. and Kritz-Silverstein, D. (1999) Bioavailable testosterone and depressed mood in older men: the Rancho Bernardo Study. *J. Clin. Endocrinol. Metab.*, **84**, 573–577.

Berkman, L.F., Seeman, T.E., Albert, M., Blazer, D., Kahn, R., Mohs, R. *et al.* (1993) High, usual and impaired functioning in community-dwelling older men and women: findings from the MacArthur Foundation Research Network on Successful Aging. *J. Clin. Epidemiol.*, **46**, 1129–1140.

Berr, C., Lafont, S., Debuire, B., Dartigues, J.F. and Baulieu, E.E. (1996) Relationships of dehydroepiandrosterone sulfate in the elderly with functional, psychological, and mental status, and short-term mortality: a French community-based study. *Proc. Natl. Acad. Sci. USA*, **93**, 13410–13415.

Bloch, M., Schmidt, P.J., Danaceau, M.A., Adams, L.F. and Rubinow, D.R. (1999) Dehydroepiandrosterone treatment of mid-life dysthymia. *Biol. Psychiatry*, **45**, 1533–1541.

Brooks-Gunn, J. and Warren, M.P. (1989) Biological and social contributions to negative affect in young adolescent girls. *Child Dev.*, **60**, 40–55.

Buckwalter, J.G., Stanczyc, F.Z., McCleary, C.A., Bluestein, B.W., Buckwalter, D.K., Rankin, K.P. *et al.* (1999) Pregnancy, the postpartum, and steroid hormones: effects on cognition and mood. *Psychoneuroendocrinology*, **24**, 69–84.

Calabrese, V.P., Isaacs, E.R. and Regelson, W. (1990) Dehydroepiandrosterone in multiple sclerosis: positive effects on the fatigue syndrome in a non-randomized study. In M. Kalimi and W. Regelson (eds), *The biologic role of dehydroepiandrosterone (DHEA)*. Berlin: W. de Gruyter, pp. 95–100.

Carlson, L.E., Sherwin, B.B. and Chertkow, H.M. (1999) Relationships between dehydroepiandrosterone sulfate (DHEAS) and cortisol (CRT) plasma levels and everyday memory in Alzheimer's disease patients compared to healthy controls. *Hormones & Behavior*, **35**, 254–263.

Casson, P.R., Santoro, N., Elkind-Hirsch, K., Carson, S.A., Hornsby, P.J., Abraham, G. *et al.* (1998) Postmenopausal dehydroepiandrosterone administration increases free insulin-like growth factor-I and decreases high-density lipoprotein: a six-month trial. *Fertil. Steril.*, **70**, 107–110.

Cawood, E.H. and Bancroft, J. (1996) Steroid hormones, the menopause, sexuality and well-being of women. *Psychol. Med.*, **26**, 925–936.

Constantino, J.N., Grosz, D., Saenger, P., Chandler, D.W., Nandi, R. and Earls, F.J. (1993) Testosterone and aggression in children. *J. Am. Acad. Child Adolesc. Psychiatry*, **32**, 1217–1222.

Cruess, D.G., Antoni, M.H., Kumar, M., Ironson, G., McCabe, P., Fernandez, J.B. *et al.* (1999) Cognitive-behavioral stress management buffers decreases in dehydroepiandrosterone sulfate (DHEA-S) and increases in cortisol/DHEA-S ratio and reduces mood disturbance and perceived stress among HIF-seropositive men. *Psychoneuroendocrinology*, **24**, 537–549.

Diamond, D.M., Branch, B.J. and Fleshner, M. (1996) The neurosteroid dehydroepiandrosterone sulfate (DHEAS) enhances hippocampal primed burst, but not long-term, potentiation. *Neurosci. Lett.*, **202**, 204–208.

Diamond, P., Brisson, G.R., Candas, B. and Peronnet, F. (1989) Trait anxiety, submaximal physical exercise and blood androgens. *Eur. J. Appl. Physiol.*, **58**, 699–704.

Dilbaz, N., Guz, H. and Arikazan, M. (1998) Comparison of serum gonadal sex hormones of early-onset schizophrenics with adult onset schizophrenics, *XXIst Collegium Internationale Neuro-Psychopharmacologicum Congress*. Glasgow, UK, Abst.# PT10010.

Erb, J.L., Kadane, J.B., Tourney, G., Mickelsen, R., Trader, D., Szabo, R. *et al.* (1981) Discrimination between schizophrenic and control subjects by means of plasma dehydro-epiandrosterone measurements. *J. Clin. Endocrinol. Metab.*, **52**, 181–186.

Fava, M., Littman, A. and Halperin, P. (1987) Neuroendocrine correlates of the Type A behavior pattern: a review and new hypothesis. *Int. J. Psychiatry Med.*, **17**, 289–307.

Fava, M., Littman, A., Lamon-Fava, S., Milani, R. Shera, D., MacLaughlin, R. *et al.* (1992) Psychological, behavioral and biochemical risk factors for coronary artery disease among American and Italian male corporate managers. *Am. J. Cardiol.*, **70**, 1412–1416.

Fava, M., Rosenbaum, J.F., MacLaughlin, R.A., Tesar, G.E., Pollack, M.H., Cohen, L.S. *et al.* (1989) Dehydroepiandrosterone-sulfate/cortisol ratio in panic disorder. *Psychiatry Res.*, **28**, 345–350.

Ferguson, H.C., Bartram, A.C.G., Fowlie, H.C., Cathro, D.M., Birchall, K. and Mitchell, F.L. (1964) A preliminary investigation of steroid excretion in depressed patients before and after electroconvulsive therapy. *Acta Endocrinol. (Copenh.)*, **47**, 58–66.

Ferrari, E., Borri, R., Casarotti, D. *et al.* (1997) Major depression in elderly patients: a chrono-neuroendocrine study. *Aging Clin. Exp. Res.*, **9**, 83.

Flynn, M.A., Weaver-Osterholtz, D., Sharpe-Timms, K.L., Allen, S. and Krause, G. (1999) Dehydroepiandrosterone replacement in aging humans. *J. Clin. Endocrinol. Metab.*, **84**, 1527–1533.

Glaser, J.L., Brind, J.L., Vogelman, J.H., Eisner, M.J., Dillbeck, M.C., Wallace, R.K. *et al.* (1992) Elevated serum dehydroepiandrosterone sulfate levels in practitioners of the Transcendental Meditation (TM) and TM-Sidhi programs. *J. Behav. Med.*, **15**, 327–341.

Goodyer, I.M., Herbert, J., Altham, P.M., Pearson, J., Secher, S.M. and Shiers, H.M. (1996) Adrenal secretion during major depression in 8- to 16-year-olds, I. Altered diurnal rhythms in salivary cortisol and dehydroepiandrosterone (DHEA) at presentation. *Psychol. Med.*, **26**, 245–256.

Goodyer, I.M., Herbert, J., Tamplin, A. and Altham, P.M.E. (2000) First-episode major depression in adolescents. Affective, cognitive and endocrine characteristics of risk status and predictors of onset. *Br. J. Psychiatry*, **176**, 142–149.

Gordon, C.M., Grace, E., Jean Emans, S., Goodman, E., Crawford, M.H. and Leboff, M.S. (1999) Changes in bone turnover markers and menstrual function after short-term oral DHEA in young women with anorexia nervosa. *J. Bone Miner. Res.*, **14**, 136–145.

Gray, A., Jackson, D.N. and McKinlay, J.B. (1991) The relation between dominance, anger, and hormones in normally aging men: results from the Massachusetts male aging study. *Psychosom. Med.*, **53**, 375–385.

Hansen, C.R., Jr., Kroll, J. and Mackenzie, T.B. (1982) Dehydroepiandrosterone and affective disorders [letter]. *Am. J. Psychiatry*, **139**, 386–387.

Harris, D.S., Wolkowitz, O.M. and Reus, V.I. (2001) Movement disorder, memory, psychotic symptoms and serum DHEA levels in schizophrenic patients, *World J. Biol. Psychiatry*, **2**, 99–102.

Heinz, A., Weingartner, H., George, D., Hommer, D., Wolkowitz, O.M. and Linnoila, M. (1999) Severity of depression in abstinent alcoholics is associated with dehydroepiandro-sterone-sulfate concentrations. *Psychiatry Res.*, **89**, 97–106.

Heuser, I., Deuschle, M., Luppa, P., Schweiger, U., Standhardt, H. and Weber, B. (1998) Increased diurnal plasma concentrations of dehydroepiandrosterone in depressed patients. *J. Clin. Endocrinol. Metab.*, **83**, 3130–3133.

Hillen, T., Lun, A., Reischies, F.M., Borchelt, M., Steinhagen-Thiessen, E. and Schaub, R.T. (2000) DHEA-S plasma levels and incidence of Alzheimer's disease. *Biol. Psychiatry*, **47**, 161–163.

Huppert, F.A., Van Niekerk, J.K. and Herbert, J. (2000) Dehydroepiandrosterone (DHEA) supplementation for cognition and well-being. *Cochrane Database Syst. Rev.*, **2**, CD00304.

Kalmijn, S., Launer, L.J., Stolk, R.P., deJong, F.H., Pols, H.A., Hofman, A. *et al.* (1998) A prospective study on cortisol, dehydroepiandrosterone sulfate, and cognitive function in the elderly. *J. Clin. Endocrinol. Metab.*, **83**, 3487–3492.

Kline, M.D. and Jaggers, E.D. (1999) Mania onset while using dehydroepiandrosterone. *Am. J. Psychiatry*, **156**, 971.

Kroboth, P.D., Salek, F.S., Stone, R.A., Bertz, R.J. and Kroboth III, F.J. (1999) Alprazolam increases DHEA concentrations in young and elderly men. *J. Clin. Psychopharmacol.*, **19**, 114–124.

Labbate, L.A., Fava, M., Oleshansky, M., Zoltec, J., Littman, A. and Harig, P. (1995) Physical fitness and perceived stress. Relationships with coronary artery disease risk factors. *Psychosomatics*, **36**, 555–560.

Leblhuber, F., Neubauer, C., Peichl, M., Reisecker, F., Steinparz, F.X., Windhager, E. *et al.* (1993): Age and sex differences of dehydroepiandrosterone sulfate (DHEAS) and cortisol (CRT) plasma levels in normal controls and Alzheimer's disease (AD). *Psychopharmacology*, **111**, 23–26.

Legrain, S., Berr, C., Frenoy, N., Gourlet, V., Debuire, B. and Baulieu, E.E. (1995) Dehydroepiandrosterone sulfate in a long-term care aged population. *Gerontology*, **41**, 343–351.

Legrain, S., Massien, C., Lahlou, N., Roger, M., Debuire, B., Diquet, B. *et al.* (2000) Dehydroepiandrosterone replacement administration: Pharmacokinetic and pharmacodynamic studies in healthy elderly subjects. *J. Clin. Endocrinol. Metab.*, **85**, 3208–3217.

Littman, A.B., Fava, M., Halperin, P., Lamon-Fava, S., Drews, F.R., Oleshanky, M.A. *et al.* (1993) Physiologic benefits of a stress reduction program for healthy middle-aged army officers. *J. Psychosom. Res.*, **37**, 345–354.

Lupien, S.J., Nair, N.P., Briere, S., Maheu, F., Tu, M.T., Lemay, M. *et al.* (1999) Increased cortisol levels and impaired cognition in human aging: implication for depression and dementia in later life. *Rev. Neurosci.*, **10**, 17–139.

Maayan, R., Yagorowksi, Y., Grupper, D., Weiss, M., Shtaif, B., Kaoud, M.A. *et al.* (2000) Basal plasma dehydroepiandrosterone sulfate level: a possible predictor for response to electroconvulsive therapy in depressed psychotic inpatients. *Biol. Psychiatry*, **48**, 693–701.

Markowitz, J.S., Carson, W.H. and Jackson, C.W. (1999) Possible dehydroepiandrosterone-induced mania. *Biol. Psychiatry*, **45**, 241–242.

McCraty, R., Barrios-Choplin, B., Rozman, D., Atkinson, M. and Watkins, A.D. (1998) The impact of a new emotional self-management program on stress, emotions, heart rate variability, DHEA and cortisol. *Integr. Physiol. Behav. Sci.*, **33**, 151–170.

Michael, A., Jenaway, A., Paykel, E.S. and Herbert, J. (2000) Altered salivary dehydroepiandrosterone levels in major depression in adults. *Biol. Psychiatry*, **48**, 989–995.

Miller, T.P., Taylor, J., Rogerson, S., Mauricio, M., Kennedy, Q., Schatzberg, A. *et al.* (1998) Cognitive and noncognitive symptoms in dementia patients: relationship to cortisol and dehydroepiandrosterone. *Int. Psychogeriatr.*, **10**, 85–96.

Morales, A.J., Haubrich, R.H., Hwang, J.Y., Asakura, H. and Yen, S.S. (1998) The effect of six months treatment with a 100 mg daily dose of dehydroepiandrosterone (DHEA) on circulating sex steroids, body composition and muscle strength in age-advanced men and women. *Clin. Endocrinol.*, **49**, 421–432.

Morales, A.J., Nolan, J.J., Nelson, J.C. and Yen, S.S. (1994) Effects of replacement dose of dehydroepiandrosterone in men and women of advancing age [published erratum appears in *J. Clin. Endocrinol. Metab.*, 1995 Sep; **80**(9),2799]. *J. Clin. Endocrinol. Metab.*, **78**, 1360–1367.

Morrison, M.F., Katz, I.R., Parmelee, P., Boyce, A.A., and Tenttave, T. (1998) Dehydroepi-androsterone sulfate (DHEA-S) and psychiatric and laboratory measures of frailty in a residential care population. *Am. J. Getriatr. Psychiatry*, **6**, 277–284.

Morrison, M.F., Redei, E., TenHave, T., Parmelee, P., Boyce, A.A., Sinha, P.S. *et al.* (2000) Dehydroepiandrosterone sulfate and psychiatric measures of frail, elderly residential care population. *Biol. Psychiatry*, **47**, 144–150.

Murialdo, G., Nobili, F., Rollero, A., Gianelli, M.V., Copello, F., Rodriguez, G. *et al.* (2000) Hippocampal perfusion and pituitary-adrenal axis in Alzheimer's disease. *Neuropsychobiology*, **42**, 51–57.

Nasman, B., Olsson, T., Backstrom, T., Eriksson, S., Grankvist, K., Viitanen, M. *et al.* (1991) Serum dehydroepiandrosterone sulfate in Alzheimer's disease and in multi-infarct dementia. *Biol. Psychiatry*, **30**, 684–690.

Oades, R.D. and Schepker, R. (1994) Serum gonadal steroid hormones in young schizophrenic patients. *Psychoneuroendocrinology*, **19**, 373–385.

Osran, H., Reist, C., Chen, C.C., Lifrak, E.T., Chicz-DeMet, A. and Parker, L.N. (1993) Adrenal androgens and cortisol in major depression. *Am. J. Psychiatr*, **150**, 806–809.

Rabkin, J.G., Ferrando, S.J., Wagner, G.J. and Rabkin, R. (2000) DHEA treatment for HIV+ patients: effects on mood, androgenic and anabolic parameters. *Psychoneuroendocrinology*, **25**, 53–68.

Ravaglia, G., Forti, P., Maioli, F., Boschi, F., Bernardi, M., Pratelli, L. *et al.* (1996) The relationship of dehydroepiandrosterone sulfate (DHEAS) to endocrine-metabolic parameters and functional status in the oldest-old. Results from an Italian study on healthy free-living over-ninety-year-olds. *J. Clin. Endocrinol. Metab.*, **81**, 1173–1178.

Reus, V.I., Wolkowitz, O.M., Roberts, E. *et al.* (1993) Dehydroepiandrosterone (DHEA) and memory in depressed patients. *Neuropsychopharmacology*, **9**, Abst. 66S.

Romeo, E., Strohle, A., Spalletta, G., di Michele, F., Hermann, B., Holsboer, F. *et al.* (1998) Effects of antidepressant treatment on neuroactive steroids in major depression. *Am. J. Psychiatry*, **155**, 910–913.

Rudman, D., Shetty, K.R. and Mattson, D.E. (1990) Plasma dehydroepiandrosterone sulfate in nursing home men. *J. Am. Geriat. Soc.*, **38**, 421–427.

Sands, D.E. and Chamberlain, G.H.A. (1952) Treatment of inadequate personality in juveniles by dehydroisoandrosterone. *Br. Med. J.*, **2**, 66.

Schneider, L.S. and Hinsey, M.S.L. (1992) Plasma dehydroepiandrosterone sulfate in Alzheimer's disease. *Biol. Psychiatry*, **31**, 205–208.

Shulman, L.H., DeRogatis, L., Spielvogel, R., Miller, J.L. and Rose, L.I. (1992) Serum androgens and depression in women with facial hirsutism. *J. Am. Acad. Dermatol.*, **27**, 178–181.

Solerte, S.B., Fioravanti, M., Schifino, N., Cuzzoni, G., Fontana, I., Vignati, G. *et al.* (1999) Dehydroepiandrosterone sulfate decreases the interleukin-2-mediated overactivity of the natural killer cell compartment in senile dementia of the Alzheimer type. *Dement. Geriatr. Cogn. Disord.*, **10**, 21–27.

Spath-Schwalbe, E., Dodt, C., Dittmann, J., Schuttler, R. and Fehm, H.L. (1990) Dehydro-epiandrosterone sulfate in Alzheimer's disease. *Lancet*, **335**, 1412.

Strauss, E.B., Sands, D.E., Robinson, A.M., Tindall, W.J. and Stevenson, W.A.H. (1952) Use of dehydroisoandrosterone in psychiatric treatment: a preliminary survey. *Br. Med. J.*, **2**, 64–66.

Stomati, M., Rubino, S., Spinett, A., Parrini, D., Luisi, S., Casarosa, E. *et al.* (1999) Endocrine, neuroendocrine and behavioral effects of oral dehydroepiandrosterone sulfate supplementation in postmenopausal women. *Gynecol. Endocrinol.*, **13**, 15–25.

Sugino, M., Ohsawa, N., Ito, T., Ishida, S., Yamasaki, H., Kimura, F. *et al.* (1998) A pilot study of dehydroepiandrosterone sulfate in myotonic dystrophy. *Neurology*, **51**, 586–589.

Sunderland, T., Merril, C.R., Harrington, M.G., Lawlor, B.A., Molchan, S.E., Martinez, R. *et al.* (1989) Reduced plasma dehydroepiandrosterone concentrations in Alzheimer's disease. *The Lancet*, **2**(8662), 570.

Takebayashi, M., Kagaya, A., Uchitomi, Y., Kugaya, A., Muraoka, M., Yokota, N. *et al.* (1998) Plasma dehydroepiandrosterone sulfate in unipolar major depression. Short communication. *J. Neural. Transm.*, **105**, 537–542.

Tode, T., Kikuchi, Y., Hirata, J., Kita, T., Nakata, H. and Nagata, I. (1999) Effect of Korean red ginseng on psychological functions in patients with severe climacteric syndromes. *Int. J. Gynaecol. Obstet.*, **67**, 169–174.

Tollefson, G.D., Haus, E., Garvey, M.J., Evans, M. and Tuason, V.B. (1990) 24 hour urinary dehydroepiandrosterone sulfate in unipolar depression treated with cognitive and/or pharmacotherapy. *Ann. Clin. Psychiatry*, **2**, 39–45.

Tourney, G. and Hatfield, L. (1972) Plasma androgens in male schizophrenics. *Arch. Gen. Psychiatry*, **27**, 753–755.

Uzunova, V., Sheline, Y., Davis, J.M., Rasmusson, A., Uzunov, D.P., Costa, E. *et al.* (1998) Increase in the cerebrospinal fluid content of neurosteroids in patients with unipolar major depression who are receiving fluoxetine or fluvoxamine. *Proc. Natl. Acad. Sci. USA*, **95**, 3239–3244.

van Goozen, S.H., Matthys, W., Cohen-Kettenis, P.T., Thijssen, J.H. and van Engeland, H. (1998) Adrenal androgens and aggression in conduct disorder prepubertal boys and normal controls. *Biol. Psychiatry*, **43**, 156–158.

van Vollenhoven, R.F., Engleman, E.G. and McGuire, J.L. (1994) An open study of dehydro-epiandrosterone in systemic lupus erythematosus. *Arthritis Rheum.*, **37**, 1305–1310.

Winterer, J., Gwirtsman, H.E., George, D.T., Kaye, W.H., Loriaux, D.L. and Cutler, G.B., Jr. (1985) Adrenocorticotropin-stimulated adrenal androgen secretion in anorexia nervosa: impaired secretion at low weight with normalization after long-term weight recovery. *J. Clin. Endocrinol. Metab.*, **61**, 693–697.

Wolf, O.T., Naumann, E., Hellhammer, D.H. and Kirschbaum, C. (1998a) Effects of dehydro-epiandrosterone replacement in elderly men on event-related potentials, memory, and well-being. *J. Gerontol. A. Biol. Sci. Med. Sci.*, **53**, M385–M390.

Wolf, O.T., Neumann, O., Hellhammer, D.H., Geiben, A.C., Strasburger, C.J., Dressendorfer, R.A. *et al.* (1997b) Effects of a two-week physiological dehydroepiandrosterone substitution on cognitive performance and well-being in healthy elderly women and men. *J. Clin. Endocr. Metab.*, **82**, 2363–2367.

Wolf, O.T., Koster, B., Kirschbaum, C., Pietrowsky, R., Kern, W., Hellhammer, D.H. *et al.* (1997a) A single administration of dehydroepiandrosterone does not enhance memory performance in young healthy adults, but immediately reduces cortisol levels. *Biol. Psychiatry*, **42**, 845–848.

Wolf, O.T., Kudielka, B.M., Hellhammer, D.H., Hellhammer, J. and Kirschbaum, C. (1998b) Opposing effects of DHEA replacement in elderly subjects on declarative memory and attention after exposure to a laboratory stressor. *Psychoneuroendocrinology*, **23**, 617–629.

Wolkowitz, O.M., Kramer, J.H., Reus, V.I. *et al.* (1999b) Dehydroepiandrosterone (NPI-34133) treatment of Alzheimer's disease: A randomized, double-blind, placebo-controlled, parallel group study [Abstract], *Annual Convention of the American Psychiatric Association*. Washington, D.C.

Wolkowitz, O.M., Reus, V.I., Keebler, A., Nelson, N., Friedland, M., Brizendine, L. *et al.* (1999a) Double-blind treatment of major depression with dehydroepiandrosterone (DHEA). *Am. J. Psychiatry*, **156**, 646–649.

Wolkowitz, O.M., Reus, V.I., Roberts, E., Manfredi, F., Chan, T., Raum, W.J. *et al.* (1997) Dehydroepiandrosterone (DHEA) treatment of depression. *Biol Psychiatry*, **41**, 311–318.

Yaffe, K. Ettinger, B., Pressman, A., Seeley, D., Whooley, M., Schaefer, C. *et al.* (1998) Neuropsychiatric function and dehydroepiandrosterone sulfate in elderly women: a prospective study. *Biol. Psychiatry*, **43**, 694–700.

Yanase, T., Fukahori, M., Taniguchi, S., Nishi, Y., Sakai, Y., Takayanagi, R. *et al.* (1996) Serum dehydroepiandrosterone (DHEA) and DHEA-sulfate (DHEA-S) in Alzheimer's disease and in cerebrovascular dementia. *Endocr. J.*, **43**, 119–123.

Zumoff, B., Walsh, B.T., Katz, J.L., Levin, J., Rosenfeld, R.S., Kream, J. *et al.* (1983) Subnormal plasma dehydroepiandrosterone to cortisol ratio in anorexia nervosa: a second hormonal parameter of ontogenic regression. *J. Clin. Endocrinol. Metab.*, **56**, 668–672.

Chapter 11

Effects of DHEA on diabetes mellitus

Kazutaka Aoki and Hisahiko Sekihara

INTRODUCTION

Dehydroepiandrosterone (DHEA) and its sulfate ester (DHEAS) are the most abundant circulating adrenal steroids in humans. The serum concentration of DHEAS in human plasma is approximately 300-fold higher than that of DHEA and 20-fold higher than that of any other steroid hormone. DHEA is considered to be a weak androgen. Peak levels of DHEA and DHEAS occur around the age of twenty and decrease gradually to 5% of these peak values by age ninety. DHEA is synthesized from pregnenolone, more than 90% of the synthesis in the total body being carried out in the adrenal gland, and DHEA is metabolized to androstenedione, testosterone, and estrogens.

The administration of DHEA to rats and mice has been reported to have beneficial effects on obesity, atheroscrelosis, osteoporosis, hyperlipidemia, and diabetes.

Non-insulin dependent diabetes mellitus (NIDDM) is characterized by impaired capacity to secrete insulin, insulin resistance, or both. Insulin resistance is considered as the result of increased glucose output in the liver and decreased glucose intake in muscle and adipose tissue. In this chapter, we describe the role of DHEA in diabetes mellitus, especially with respect to insulin resistance as conducted in our laboratory.

EFFECT OF DHEA ON DB/DB MICE

C57BL/KsJ-db/db mice, which become obese, hyperglycemic, and hyperinsulinemic (Gunnarsson, 1975), represent a typical model of NIDDM. Coleman *et al.* (1984) reported that dietary administration of DHEA to db/db mice induced remission of hyperglycemia and increased insulin sensitivity (Coleman *et al.*, 1982, 1984). Many other studies suggest that DHEA increases sensitivity to insulin (Ladriere *et al.*, 1997; Kimura *et al.*, 1998; Mukasa *et al.*, 1998). These studies do not state the mechanism of increasing insulin sensitivity fully when DHEA was administered to animal models. Therefore, we evaluated glucose metabolizing enzyme activities in the liver and muscle in db/db mice.

Plasma insulin levels of db/db mice were higher than that of db/+m mice as shown in Table 11.1. Despite hyperinsulinemia, the activities of hepatic

Table 11.1 Plasma insulin, BG (blood glucose), and body weight in six groups of mice

	db/+m	db/+m+DHEA	db/+m+Tro	db/db	db/db+DHEA	db/db+Tro
Insulin (pg/ml)	979±51	910±72	1120±210	4634±309*	11449±2005#	3234±260
BG (mg/dl)	159±2	145±2	131±4*	557±5*	349±9#	324±9#
Initial body weight (g)	27.8±1.0	27.1±0.4	28.3±1.7	48.1±1.1	49.2±0.8	49.4±1.2
Final body weight (g)	29.3±1.8	29.0±0.4	30.6±3.6	55.7±0.2*	55.1±0.8	59.8±1.5#

Notes

Each value is the mean±SEM. $*P < 0.05$ vs. db/+m mice and $^{#}P < 0.05$ vs. db/db mice. db/+m, db/+m mice treated with standard food; db/+m + DHEA, db/+m mice treated with DHEA; db/+m +Tro, db/+m mice treated with troglitazone; db/db, db/db mice treated with standard food; db/db+DHEA, db/db mice treated with DHEA ($n = 5$); db/db + Tro, db/db mice treated with troglitazone. From Aoki K. *et al.* (1999), *Diabetes*, **48**, 1579.

glucose-6-phosphatase (G6Pase) and fructose-1,6-bisphosphatase (FBPase), which are gluconeogenic enzymes and are normally suppressed by the action of insulin (Pilkis and Granner, 1992), were increased in db/db mice when compared to db/+m mice (Aoki *et al.*, 1999) as shown in Figure 11.1. The increased activities of G6Pase and FBPase are considered to be important in the development of hyperglycemia in db/db mice. Dietary administration of DHEA significantly decreased blood glucose in db/db mice and hepatic G6Pase and FBPase activities both in db/db and db/+m mice (Figure 11.1). Troglitazone, which was the first drug to be introduced in clinical medicine to improve insulin sensitivity, also decreased blood glucose and hepatic G6Pase and FBPase activities both in db/db and db/+m mice. As shown in Figure 11.2, hepatic G6Pase and FBPase activities exhibit a linear relationship with blood glucose in all the groups of mice, thus suggesting that the activities of G6Pase and FBPase are closely related to blood glucose levels.

However, there are no significant positive relationships between muscle enzyme activities and blood glucose levels.

Since androstenedione, a DHEA metabolite, exhibited almost no effect on either of these enzyme activities or blood glucose in db/db mice, actions of DHEA, which are similar to troglitazone, are presumed to be caused by DHEA itself. DHEA is considered to be a modulating agent for the activities of hepatic gluconeogenic enzymes in db/db mice.

DHEA is considered to increase the insulin sensitivity in the same way as troglitazone does in the liver; however, the effects are different for DHEA and troglitazone as shown in Table 11.1. Administration of troglitazone increased body weight in db/db mice; however, DHEA did not cause a change in body weight. In another study, administration of DHEA decreased the body weight of diabetic rats (Pilkis and Granner, 1992). Increase in plasma insulin by DHEA in db/db mice was considered to be due to prevention of islet cell atrophy, as previously reported histologically by Coleman *et al.* (1982). However, the plasma insulin levels in troglitazone-treated db/db mice were not changed. From this viewpoint, DHEA has beneficial effects in diabetes mellitus that are different from those of troglitazone.

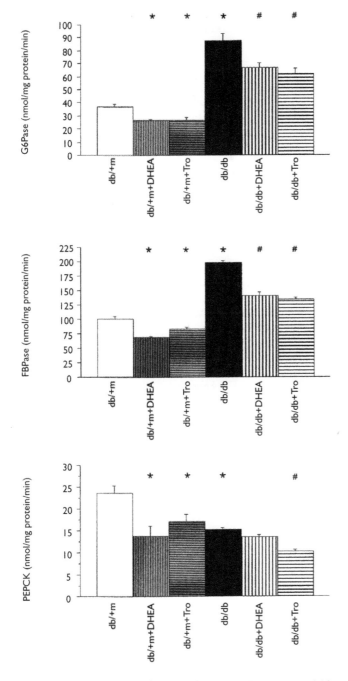

Figure 11.1. Comparison of hepatic gluconeogenic enzyme activities among six groups of mice. Tro represents troglitazone. Each column and bar represents mean ±SE. *P < 0.05 vs. control db/+m mice and #P < 0.05 vs. control db/db mice. From Aoki, K. *et al.* (1999) *Diabetes,* **48,** 1579–1585.

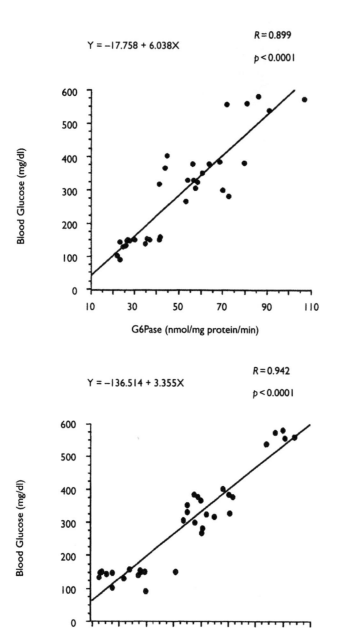

Figure 11.2 Regression analysis between hepatic G6Pase or FBPase activities and mean blood glucose for the last five days. From Aoki, K. *et al.* (1999) *Diabetes*, **48**, 1579–1585.

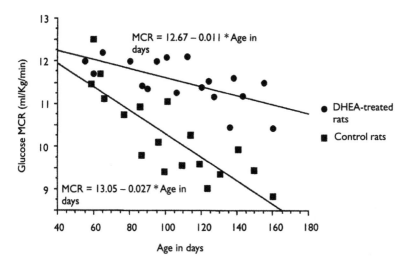

Figure 11.3 Glucose metabolic clearance rate (MCR) of control and DHEA-treated rats. From Mukasa, K. *et al.* (1998) *J. Steroid Biochem. Mol. Biol.*, **67**, 355–358.

EFFECT OF DHEA ON INSULIN SENSITIVITY IN OLDER RATS (MUKASA *ET AL.*, 1998)

It has already been reported that glucose sensitivity (tolerance) decreases with age and glucose intolerance develops as a part of the ageing process (Fink *et al.*, 1983). Therefore, we evaluated the effects of exogenous DHEA on tissue sensitivity of insulin that are associated with ageing in rats using the hyperinsulinemic euglycemic clamp technique. The glucose clamp technique is commonly used to assess insulin sensitivity. This technique is considered to show the insulin sensitivity mainly in muscles and adipose tissue. The glucose metabolic clearance rate (MCR) of control rats showed a gradual decline with advancing age. The glucose MCR of the DHEA-treated rats also exhibited a gradual decline with the ageing process. However, the MCR of the DHEA-treated rats was significantly higher than that of the control rats as shown in Figure 11.3. Since glucose MCR is a parameter indicating the insulin sensitivity especially in muscles and body composition, and was not changed after the injection of DHEA, DHEA is considered to work on muscles by increasing insulin sensitivity.

EFFECT OF DHEA ON SERUM TUMOR NECROSIS FACTOR ALPHA (TNF-α) IN GENETICALLY OBESE ZUCKER FATTY RATS (KIMURA *ET AL.*, 1998)

In this study, the effects of DHEA treatment on insulin sensitivity were investigated in genetically obese Zucker rats, an animal model for insulin resistance, using the euglycemic clamp technique. After administration of 0.4% DHEA to

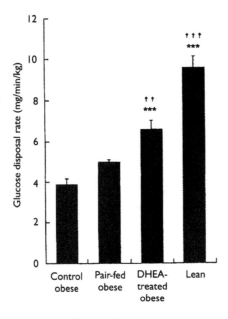

Figure 11.4 Effects of DHEA treatment on glucose disposal rate. Results are presented as mean ± SE. ***P < 0.01 *vs.* control obese rats; ††P < 0.01; †††P < 0.001 *vs.* pair-fed obese rats. From Kimura, M. *et al.* (1998) *Endocrinology,* **139**, 3249–3253.

female obese Zucker rats for ten days, body weight and plasma insulin decreased and the MCR (reflecting glucose disposal rate), which was normally reduced in obese rats, increased significantly when compared with the control obese rats as shown in Figure 11.4. It is known that administraton of DHEA to rat decreases food intake. The pair-fed obese rats also showed levels of weight reduction similar to those of DHEA-treated rats, the increase in MCR of the DHEA-treated rats was significantly greater than that of the pair-fed rats, suggesting a direct ameliorating effect of DHEA on the insulin sensitivity of obese rats. TNF-α is a cytokine that is secreted from adipose tissue and causes insulin resistance. Serum TNF-α levels were measured with a bioassay that quantifies the TNF-α level based on its cytotoxicity to LM DEFINE cells. The serum TNF-α level was also reduced signific-

Table 11.2 Effect of DHEA treatment on serum TNF-α (mU/ml)

Sample number	1	2	3	4	5	6	7	8	9	10
Control obese	252.3	32.2	634.6	257.7	242.5	26.2	304.5	164.6	ND	ND
Pair-fed obese	160.1	304.1	252.3	85.3	304.6	382.5	158.4	ND	ND	ND
DHEA-treated obese	ND	ND	ND	ND	ND	ND	ND	ND	ND	ND
Lean	ND	ND	ND	ND	ND	ND	ND	ND	ND	ND

Note
ND, not detectable. From Kimura, M. *et al.* (1998) *Endocrinology,* **139**, 3249.

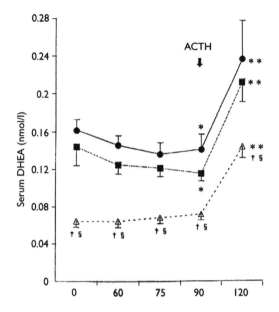

Figure 11.5 The changes in serum DHEA levels in control, NI and HI groups during the hyper-insulinemic-euglycemic clamp. Value are mean±SE. ●, Control; ■, NI; Δ, HI. $^{\dagger}P < 0.05$ vs. control; $^{\S}P < 0.05$ vs. NI; $^{*}P < 0.05$ vs. baseline; $^{**}P < 0.01$ vs. baseline. From Yamaguchi, Y. *et al.* (1998) *Clin. Endocrinol.*, **49**, 377–383.

antly in DHEA-treated mice; however, it was not reduced in pair-fed obese rats as indicated in Table 11.2. These results suggest that DHEA treatment reduces body weight and serum TNF-α independently, and that both may ameliorate insulin resistance in this rat model.

SERUM DHEA LEVELS IN DIABETIC PATIENTS WITH HYPERINSULINEMIA (YAMAGUCHI *ET AL.*, 1998)

In this study, we evaluated the relationships between the serum DHEA levels and insulin resistance in humans. Nestler *et al.* (1987) reported that a progressive decline of serum DHEAS levels during hyperinsulinemic-euglycemic clamp in normal men and women did not affect serum testosterone, progesterone, or cortisol levels (Nestler *et al.*, 1987). Furthermore, they demonstrated that a reduction in serum insulin concentration was associated with an increase in serum DHEA and DHEAS (Nestler *et al.*, 1995). In our study, to elucidate further the interaction between insulin and DHEA concentrations, we evaluated the serum DHEA and DHEAS levels in diabetic patients with hyperinsulinemia and their changes during ACTH stimulation and hyperinsulinemic-euglycemic clamp as shown in Figure 11.5. Twenty-four subjects with NIDDM, including 12 hyperinsulinemic (HI) subjects (fasting serum insulin concentrations at 10 mU/ml (71.8 pmol/l)), and 12 non-hyperinsulinemic (NI) subjects, and 10 normal control subjects were studied. Their serum DHEA

Table 11.3 Serum DHEA, DHEA-S, cortisol and ACTH levels

	Control	NI	HI
DHEA (nmol/l)	0.16±0.01	0.14±0.02	0.06±0.006*##
DHEA-S (nmol/l)	46.7±4.2	38.9±7.0	26.8±3.4*##
Cortisol (nmol/l)	304.6±44.6	279.1±22.2	266.9±13.5
ACTH (nmol/l)	1.09±0.12	1.32±0.13	1.16±0.13

Notes
Values are mean±SEM. *$P < 0.05$ vs. control; #$P < 0.05$ vs. NI (10). From Yamaguchi, Y.
et al. (1998) Clin. Endocrinol. (Oxf), **49**, 377

levels were compared during hyperinsulinemic-euglycemic clamp and after ACTH stimulation. Diabetic patients with hyperinsulinemia produced significantly lower serum DHEA and DHEAS levels than did the controls (Table 11.3). After ACTH stimulation, these patients also produced significantly lower DHEA levels. During the hyperinsulinemic-euglycemic clamp, serum DHEA concentrations of diabetic patients with hyperinsulinemia remained low and did not decline further, although those of the control subjects and of the non-hyperinsulinemic diabetic patients showed a significant decline in serum DHEA levels as shown in Figure 11.5. Even after ACTH stimulation during the clamp, the serum DHEA level in hyperinsuline-mic patients was still significantly lower than in the controls. These studies suggest that in diabetic patients with hyperinsulinemia, the DHEA level baseline is chronic-ally and maximally suppressed when compared to that of the control subjects or of the non-hyperinsulinemic diabetic patients, and thus not decreased further by exogenous insulin infusion during hyperinsulinemic-euglycemic clamp.

SUMMARY

We described the relationships between DHEA and insulin resistance based on the diabetic animal models and human studies. DHEA is considered to increase insulin sensitivity due to the effects in the liver, muscle, and adipose tissue in animal models. In humans, the serum DHEA concentration is associated with hyperinsulinemia in diabetes. Its role and side effects in the human body have yet to be elucidated. We think that exogenous DHEA administration is not currently warranted in diabetic patients. However, as DHEA has many beneficial effects, we should consider the possible use of DHEA as an endogeneous insulin sensitizing agent in future appli-cations, and the details of the mechanisms of DHEA actions should be evaluated in further studies.

REFERENCES

Aoki, K., Saito, T., Satoh, S., Mukasa, K., Kaneshiro, M., Kawasaki, S. et al. (1999) Dehydro-epiandrosterone suppresses the elevated hepatic glucose-6-phosphatase and fructose-1,6-bisphosphatase activities in C57BL/KsJ-db/db mice: comparison with troglitazone. Diabetes, **48**, 1579–1585.

Coleman, D.L., Leiter, E.H. and Schwizer, R.W. (1982) Therapeutic effects of dehydro-epiandrosterone (DHEA) in diabetic mice. *Diabetes*, **31**, 830–833.

Coleman, D.L., Schwizer, R.W. and Leiter, E.H. (1984) Effect of genetic background on the therapeutic effects of dehydroepiandrosterone (DHEA) in diabetes-obesity mutants and aged normal mice. *Diabetes*, **33**, 26–32.

Fink, R.I., Kolterman, O.G., Griffin, J. and Olefsky, J.M. (1983) Mechanisms of insulin resistance in aging. *J. Clin. Invest.*, **71**, 1523–1535.

Gunnarsson, R. (1975) Function of the pancreatic B-Cell during the development of hyper-glycemia in mice homozygous for the mutations "diabetes" (db) and "misty" (m). *Diabetologia*, **11**, 431–438.

Kimura, M., Tanaka, S., Yamada, Y., Kikuchi, Y., Yamakawa, T. and Sekihara, H. (1998) Dehydroepiandrosterone decreases serum TNF-α and restores insulin sensitivity: independent effect from secondary weight reduction in genetically obese Zucker fatty rats. *Endocrinology*, **139**, 3249–3253.

Ladriere, L., Laghmich, A., Malaisse-Lagae, F. and Malaisse, W.J. (1997) Effect of dehydro-epiandrosterone in hereditarily diabetic rats. *Cell. Biochem. Funct.*, **15**, 287–292.

Mukasa, K., Kaneshiro, M., Aoki, K., Okamura, J., Saito, T., Satoh, S. *et al.* (1998) Dehydroepiandrosterone (DHEA) ameliorates the insulin sensitivity in older rats. *J. Steroid Biochem. Mol. Biol.*, **67**, 355–358.

Nestler, J.E., Beer, N.A., Jakubowicz, D.J., Colombo, C. and Beer, R.M. (1995) Effects of insulin reduction with benfluorex on serum dehydroepiandrosterone (DHEA), DHEA-sulfate, and blood pressure in hypertensive middle-aged and elderly men. *J. Clin. Endo-crinol. Metab.*, **80**, 700–706.

Nestler, J.E., Clore, J.N., Strauss, III J.F. and Blackard, W.G. (1987) The effects of hyper-insulinemia on serum testosterone, progesterone, dehydroepiandrosterone sulfate, and cortisol levels in normal women and in a woman with hyperandrogenism, insulin resistance, and acanthosis nigricans. *J. Clin. Endocrinol. Metab.*, **64**, 180–184.

Pilkis, S.J. and Granner, D.K. (1992) Molecular physiology of the regulation of hepatic gluconeogenesis and glycolysis. *Annu. Rev. Physiol.*, **54**, 885–909.

Yamaguchi, Y., Tanaka, S., Yamakawa, T., Kimura, M., Ukawa, K., Yamada, Y. *et al.* (1998) Reduced serum dehydroepiandrosterone levels in diabetic patients with hyperinsulinemia. *Clin. Endocrinol.*, **49**, 377–383.

Chapter 12

The clinician's view of DHEA: an overview

Frank Svec

How do active clinicians get authoritative information on new therapies? Conventional wisdom suggests they consult with local specialists, read the latest editions of medical textbooks and review articles from current professional journals. Occasionally, information on drugs is delivered by representatives of pharmaceutical houses. These avenues work well when dealing with conventional drugs, but how well do they work when a clinician is trying to educate himself about a novel therapy that is not yet accepted by the established medical community?

The use of dehydroepiandrosterone (DHEA) in a busy, general practice may be a case in point. At this time, as there are no recognized indications for this steroid in the treatment of any disorder, one might wonder why a clinician would ever have to learn about DHEA. Most likely it would be because one of his patients asked whether they should use DHEA as a health supplement. Increasingly, individuals are supplementing their physician's therapies with novel diets, food additives, vitamins and minerals as a means of maintaining health and avoiding disease. In the United States, this situation may be the most developed because commercial sources of these agents are allowed to promote directly to consumers. However, this problem is destined to spread to other countries. Because of the ubiquity of the internet, and the difficulty of monitoring shipments of packages to individuals, these products are available to consumers worldwide. The issues seen in the United States may just be the opening chapter of a more global story.

Most companies aggressively promote their supplements as safe and effective. Occasionally, these commercial sources suggest to consumers that they obtain the advice of a physician before starting their products. More often, there is no recommendation that a physician be consulted. Still, some individuals are concerned enough about the possibility of untoward side effects or the possibility that supplements may interfere with the actions of other medications they are taking that they seek the advice of their physician before embarking on years of exogenous therapy. How would a clinician faced with a patient who wants advice about DHEA learn about its clinical utility? It is doubtful he would have learned anything about this agent during his training. No representative of a pharmaceutical company will inform him about the nuances of therapy as none promotes its use. Few specialists use it. If a clinician needed to know more about this steroid, how would he learn about the utility of DHEA? It is a subject that he would have to evaluate himself. That leads to a number of questions: What *would* he learn? What would be his impression from such literature research? How useful is DHEA in a general practice? How balanced is the information?

This chapter presents one view of what a clinician might find if he tries to learn about DHEA. The information is presented in the order a physician might employ at the end of a busy day. The chapter begins with what one would learn if he were to turn first to some traditional major textbooks that may be found in his office library. Next, the information that can be gleaned through the use of a computer and the Internet will be evaluated; currently more and more professionals (and lay people) are turning to this source for seemingly unlimited information. Finally, some of the articles available about DHEA in the most current issues of medical journals will be reviewed to discover what a clinician might find if he scanned the professional literature.

TEXTBOOKS

Most physicians keep a library in their office to which they can turn when faced with the need to understand a new therapeutic agent. Cecil's Textbook of Medicine is a likely place to start the search for information on DHEA (Loriaux, 1996). A quick scan of the index shows it is mentioned under the heading of the adrenal gland. That chapter points out that adrenal steroids act through cellular receptors and the author notes that there are no known receptors for DHEA. He eventually makes the statement that DHEA, "...has no identifiable biologic action." Certainly from this authoritative source a clinician would not advocate that his patients start using this steroid.

Other major texts offer comparably negative or vague discussions about any beneficial role that DHEA might play in health maintenance. Fitzgerald (1997), in *Current Medical Diagnosis and Treatment*, writes that, "...the adrenal continues to make large amounts of dehydroepiandrosterone sulfate (DHEAS) and dehydroepiandrosterone (DHEA), which have no known significance during adult life, having minimal androgenic activity; but they continue to be the adrenals' most abundantly secreted steroids." He tantalizes the reader by adding, "...for unknown reasons – there is a positive correlation between DHEAS level and longevity." Any curious reader may find this brief statement of more than passing interest. Why would this steroid be related to lifespan? There is no further information, unfortunately, in that text. Hazzard *et al.* (1990) in *Principles of Geriatric Medicine and Gerontology* make no note of this possible effect in their 1990 edition.

Williams and Dluhy (1998) in *Harrison's Textbook of Internal Medicine* also offer only a few words on DHEA and they imply that it is an inactive steroid. If it has any action at all, it is only through its conversion to potent androgens and estrogens in extraglandular tissue. Still, these authors subtly add to their comments that, "DHEA also has poorly understood effects on the immune and cardiovascular systems." What is this brief comment all about? Could this be the way DHEA is linked to longevity? Could these off-hand comments hold a key to some unexplored salutary effects of DHEA?

What would a clinician conclude from this brief search of DHEA in the established treatises of medical literature? That DHEA is of no current therapeutic significance. Texts provide only scant space to this steroid. The comments are susually presented in those sections on adrenal physiology, not on therapeutics. On the other hand, one might infer that there is something worth pursuing in this area, for some of the leading authorities offer cryptic comments that DHEA may be related to life span, have effects on the cardiovascular system and may modulate the immune response. These statements might pique the curiosity of many a physician; there may be something to his patient's questions about the potential application of DHEA in preserving health. Where would such a curious physician dig for more information?

INTERNET

Increasingly, clinicians have computers in their workplaces that are connected to the World Wide Web. Because of the speed and versatility offered by a computerized search, many physicians, especially younger ones, employ this means of searching for up to date information. Such a search would surprise the inquiring physician, for unlike the review of textbook material that turned up only a trace of information, with the computer, the problem turns out to be information overload.

In March 2000, when the word "DHEA" was entered into each of six commonly used search engines available in the United States a cornucopia of "hits" was gathered. Table 12.1 shows the number of references returned by each site. The values range from a manageable 19 to an indigestible 12,000. What does one learn when some of these sites are opened?

Some are designed to appear authoritative and independent of any sales effort, delivering "unbiased" information. Others are clearly commercial, geared toward selling products. One site that was found by several search engines was termed the Dehydroepiandrosterone home page whose address is http://www.naples.net/~nfn03605/. This title certainly suggests the physician is on target for learning more about DHEA. Indeed, the search engine Google.com, which found nearly 12,000 references when asked to search the term DHEA, found this site when asked to find *the* single most relevant site. At first, this webpage appears to be a compendium of web links to information about DHEA, melatonin and testosterone. Review of this index dizzies a clinician with the wealth of information available on

Table 12.1 Number of sites found using different search engines. In March 2,000 the abbreviation "DHEA" was placed into the search function of the six different engines listed. The number of references reported is listed

Site	References
Google.com	11,896
GoTo.com	53
Hotbot.com	>5,000
Lycos.com	10,875
Netscape.com	19
Yahoo.com	36

DHEA and its potential role in health and disease. According to this site, DHEA is linked to human evolution, AIDS, sudden infant death, epilepsy, homosexuality, chronic fatigue syndrome, obesity and ageing, just to name a few. Clicking onto these links shows that nearly all lead to treatises by James Michael Howard of Fayetteville, Arkansas, USA. His credentials, if any, for editing and writing these articles are not given. Thus, most probably, this site may be safely put to the side even though it was found so frequently.

Other sites on the Web offer what appears to be more genuine information. One is http://www.natrol.com/products/dhea-wp.html. This site is a "DHEA White Paper", which is maintained by Natrol, Inc, a company that manufactures and distributes dietary supplements. In small type at the end of the article it notes that the information is prepared by a law firm which specializes in "food, drug and advertising law". The site contains some useful information on the legal issues that surround DHEA and may prove important for the clinician to peruse. They note that DHEA, according to the Drug Enforcement Administration, is not an anabolic steroid and thus not a controlled substance. They also point out, on the other hand, that it is a "dietary supplement" and give a definition of this term, "a dietary substance for use by man to supplement the diet by increasing the total dietary intake." This circular definition is followed by an important statement that is much clearer, "if therapeutic claims are made for DHEA in product labeling or advertising, it could be regulated as a 'drug' pursuant to the Federal Food, Drug and Cosmetic Act. The term 'drug' is defined in Section 201(g) of the Act to include articles intended for use in the diagnosis, cure, mitigation, treatment or prevention of a disease in man or other animals. Thus, if claims are made in DHEA labeling recommending its use for treatment or prevention of a disease, the US Food and Drug Administration (FDA) could regulate the product as a drug."

This lengthy statement clarifies many of the oblique comments the clinician finds while evaluating the lay literature about DHEA. Although the product is promoted by commercial enterprises as being good for health, they do not promote it as having any specific effect. All claims are couched in generalities that suggest *potential* effects that may be "associated" with benefit, without actually claiming that the supplement does anything.

Companies face a conundrum; how do they promote a substance without claiming it is useful. Many sites utilize ingenious ways of skirting the problem. For example, they first present selections from authentic scientific studies published in legitimate journals that show DHEA has beneficial effects in experimental animals and humans and then juxtapose these reports with advertising of their product. They allow the reader to make the obvious connection.

Some sites are less restrained in promoting the potential beneficial effects of DHEA. UltraNutrition (http://www.ultranutrition.com/facts.htm) starts out with the statement that DHEA may be, "Perhaps the breakthrough of the century...and generates a very wide variety of health and longevity benefits." The site goes on to quote scientific studies such as the one from the 1995 meeting of the New York Academy of Sciences where they claim, it was announced that older men treated with DHEA had a "dramatic improvement in their ratio of lean mass to body fat."

Inspection of additional web sites uncovers a drumbeat of positive findings facing the inquisitive clinician. Worldwide Labs' DHEA Central (http://www.

webadprod.com/dhea/dheafaq.htm) summarizes "DHEA's capacity to control all obesity states." Further, it claims that Arthur Schwartz found scientific evidence that DHEA prolongs the life span of animals as much as 24–36 months. No references were given for these claims, so a clinician might be excused for not following up on these assertions immediately. Falling DHEA levels were said to lead a person to be prone to cancer, AIDS, heart disease, diabetes, obesity, high blood pressure and Parkinson's disease. (It may be important to note here that this same group promotes the use of Laetrile.) Following the lead that DHEA may alter ageing, http://www.cockatoo.com/dhea/claims that, "the effects of DHEA can be summarized as simply rejuvenating" and then adds that sex drive is improved in both men and women with DHEA treatment.

Besides these sites written by either anonymous or lay authors, there are others allegedly written by medical experts. Two such sites are http://www.raysahelian.com/dhea.html and http://www.ceri.com/dhea.htm. Each is written by a medical doctor. The first site summarizes the findings of a book sold by Ray Sahelian, MD and promotes the beneficial effects of DHEA mostly by extolling the positive findings of the paper by Morales and Yen. To add even further credence to his writings, this author includes a transcript of a conversation he had with Dr Seymour Lieberman of the St Luke's-Roosevelt Hospital Center in New York. (Incidentally, a search of the medical literature using Medline did not turn up any articles authored by Dr Sahelian). The second site is written by Ward Dean, MD, but his credentials were not provided. In this report Dr Dean extolls the benefits of DHEA in preventing cancer, obesity, abnormal glucose metabolism, excessive appetite, ageing as well as deterioration of brain function and immune function. He provides references to the scientific literature for many of these assertions.

Besides all the beneficial effects, a few sites assure potential clients that side effects are few and benign. For example, http://www.ultranutrition.com/facts.htm notes that, "World-class ageing athletes and body-builders are known to far exceed the usual dosage of 50 to 100 mg per day...the benefits are considerable, with virtually no known side effects." http://www.webadprod.com/dhea/dheafaq.htm uses the newspaper, *USA Today*, as their reference to the safety of DHEA and states, "According to *USA Today*, short-term use at moderate doses – 25–50 mg a day – should pose no problem. There's a right dose for every medication. When the dose is exceeded, side effects occur. It's the same with DHEA, when low doses are used, people feel well-being." In fairness, they do delineate some side effects like oiliness of skin, pimples, facial hair, irritability and hair loss.

These expositions on safety might bring the clinician to consider the purity and dosage of commercial DHEA. Sites brim with information. Natrol states, the DHEA they sell is synthetically manufactured and is approximately 99% pure. Worldwide Labs claims that their DHEA is of the highest quality and greatest purity available. The tablets are pharmaceutical grade. The suspicious reader might note that no techniques of analysis are referenced, nor are there any independent analyses referenced to confirm these statements. With regards to dosage, most sites recommend 25–50 mg per day. Worldwide Labs probably offers the most detailed information: They note that DHEA should be administered in the morning, with a meal and in the beginning a person should start with a dose of 5–10 mg per day.

What would a clinician's final impression of DHEA be after "surfing" to such sites? The naive physician may, like his lay patient, be impressed by the seemingly unending rainbow of beneficial effects attributed to the hormone. Although the claims appear somewhat exaggerated, the references to legitimate articles tend to substantiate them in bold outline. Some sites even offer suggestions as to why such a wonderful product is not sold by pharmaceutical houses. "In the United States, DHEA must first be approved as a drug by the FDA before it can be marketed for medical purposes. Unfortunately, this is an adversarial process (the drug companies advocating for the drug and the FDA demanding proof of efficacy and safety) which takes up to 100 million dollars and a decade to accomplish. Without a patent to restrict competition, prices cannot be raised high enough to recover the investment in the approval process. DHEA is an unpatentable substance (www.ceri.com/dhea.htm)." The inference might be that nefarious forces of economics are keeping this wonderful additive off the market.

MEDICAL JOURNALS

Hopefully, after finding so little information about DHEA in medical textbooks and so many overblown promises about its benefits while "surfing the web", a curious and careful practitioner would turn to the medical journals. Here he would find ample information that is assuredly both reliable and rigorous. However, would a neophyte be any less confused after reading these articles?

An electronic search of the literature under the heading of DHEA reveals nearly 50 articles published during the first six months of the year 2000. They are in well respected journals and conducted by some of the finest research groups around the world. Papers come from Japan, Europe and the United States. The geographic dispersion of the research seemingly assures the clinician that there must be something very significant going on with DHEA.

In general, these 50 or so articles can be grouped into three categories: (1) Those that involve giving DHEA to experimental animals; (2) those that involve correlations between the serum levels of DHEA and human disease states; and, (3) those that involve the administration of DHEA to humans. Even a brief perusal of these select, recent articles would lead a clinician to be interested in the medical potential of DHEA.

The animal studies provide some of the most exciting results. To begin with, Hayashi *et al.* (2000) in Japan, evaluate the effect of exogenous DHEA on atherosclerosis development in cholesterol-fed rabbits. Feeding such animals DHEA for 10 weeks leads to a 60% decrease in atherosclerotic lesions. Additionally, they showed that the improvement induced by DHEA correlates with an increased production of nitric oxide by vascular tissue. Similar findings were induced by estradiol. As the effect of DHEA was blocked by an antiestrogen, these researchers concluded "that appoximately 50% of the total antiatherosclerotic effect of DHEA was achieved through the conversion of DHEA to estrogen. NO may also play a role in the antiatherosclerotic effect of DHEA and 17 beta-estradiol." Clearly, these results are promising, current and noteworthy.

Beneficial cardiovascular effects of DHEA in rats are also reported by Jarrar *et al.* (2000) in the United States, using a model of trauma-hemorrhage. Rats were laparotomized and then bled to maintain a reduced arterial pressure. The goal was to simulate trauma with hypoperfusion. They found that cardiac output was diminished in the experimental animals, but the administration of DHEA after resuscitation returned output to control values. Liver function was likewise protected by DHEA. Again the simultaneous administration of an estrogen receptor antagonist blocked the effect of DHEA suggesting that it must be converted to an estrogen before it is active. The group concludes that "DHEA is a safe and inexpensive adjunct to fluid resuscitation for restoring the depressed cardiac and hepatocellular responses after severe hemorrhagic shock in male subjects."

Besides these optimistic reports of the effects of DHEA on the circulatory system, there are comparable findings on the response to stress. Kalimi's group, which has long been interested in DHEA, reported that the administration of just 5 mg of DHEA, attenuates the adverse effects of chronic stress in rats (Hu *et al.*, 2000). Possibly an important clue to the mechanism of DHEA's action that they report is that levels of free radicals are diminished with DHEA.

Lardy's group at the University of Wisconsin reports that the administration of either DHEA or its 7-oxo derivative improves the memory of young mice treated with scopalamine and that the latter improved the memory of older mice (Shi *et al.*, 2000). They suggest that, "The possible effect of 7-oxo-DHEA in human memory problems deserves investigation." A tantalizing report from a Georgetown University group showing that DHEA is synthesized in cells derived from human brain, adds further encouragement to this potential area (Brown *et al.* 2000).

These papers alone should stimulate the curiosity of any practitioner. However, perusal of just a few of the papers from the second category (those that correlate levels of serum DHEA and human disease processes) would intrigue him even more. Further, unlike the over-hyped reports found on the World Wide Web, these studies are the fruit of highly regarded laboratories, subjected to peer-review before publication and the whereabouts of all the scientists involved in the study identified.

Many of the most promising reports evaluate the relationship between DHEA and immune function. For example, a group at the NIH reports that the level of DHEA was lower in patients newly afflicted with synovitis/rheumatoid arthritis than appropriate controls (Kanik *et al.*, 2000). There was no relationship between indices of inflammation and either ACTH or cortisol, but there was an inverse correlation between these parameters and DHEA. They also report that the level of DHEA decreases with age more dramatically in the affected group than in controls. A German group presents a similarly positive report in a cohort of patients with systemic lupus erythematosus (SLE) (Zietz *et al.*, 2000). They found that DHEA levels were lower in those with disease as compared to healthy controls.

Immune diseases are not the only ones that show a correlation with DHEA. Reiter *et al.* (2000) in Austria, report that the mean level of DHEA was lower in those with erectile dysfunction who were less than 60 years of age than in controls. Again a comparison between the serum concentration of DHEA and age showed a different curve in the affected group as compared to controls.

DHEA has frequently been related to cardiovascular diseases and the reports in the medical literature during the early part of 2000 are no different. Moriyama *et al.* (2000) report that the level of DHEAS was decreased in a cohort of Japanese who had chronic heart failure and the magnitude of the decrement correlated with the severity of their disease. They made another important observation when they found that DHEAS levels were inversely correlated with oxidative stress.

The final area that would interest a clinician deals with patients who have neurologic/psychiatric disorders. A group from the Department of Psychiatry in Cambridge reports, there was an association between DHEA levels and negative mood in a group of adolescents that went on to have a major depression during the ensuing year (Goodyer *et al.*, 2000). This report is extremely valuable because it is prospective. Likewise, Hillen *et al.* (2000) in Germany report prospective data on the relationship between Alzheimer's disease and DHEAS levels. They found that the DHEAS level in a group of 14 individuals who developed Alzheimer's within three years of the initial blood sample were lower than those of two control groups that were from the same population. One control group was matched for age and coexisting morbidities and the other was matched for age but free of comorbidities. Certainly, any clinician who had seen this report as well as the above-mentioned papers that DHEA is made in human brain cells and that 7-oxo-DHEA improves memory in old mice would have surmised that DHEA may be important in neurodegeneration.

Amongst all these glowing reports a clinician may note a few papers that cast some shadows over the area. First, although one German group reports that DHEAS was low amongst those who went on to develop Alzheimer's, a group from the University of Pennsylvania reports an anomalous positive correlation between DHEAS levels and cognitive dysfunction in a cohort of frail elderly subjects (Morrison *et al.*, 2000). Higher levels of DHEAS predicted worse functioning. Similarly, although there are many favorable reports in humans about the relationship between DHEA and cardiac disease, and a plethora of suggestive studies in experimental animals, Kiechl *et al.* (2000) report that in 867 humans between the age of 40 and 79 enrolled in a prospective health study, there was no correlation between DHEAS and atherosclerosis. Although disconcerting, these studies might be accepted if there were studies showing that the administration of DHEA to select humans was beneficial.

It is at this point when the clinician scans the literature for the effects of exogenous DHEA on humans that he may turn suspect about the beneficial effects of DHEA, for although there are many reports on the positive effects of DHEA in experimental animals, and dozens of reports correlating low levels of DHEA to a variety of human diseases, there are precious few studies in which the administration of DHEA to humans resulted in beneficial findings. For example, in the first part of 2000 Baulieu *et al.* (2000) reported on the administration of 50 mg of DHEA for a year to subjects who were between 60 and 79 years old. The protocol was double blinded and prospective. No significant adverse effects were noted in this group of 280 volunteers, however, at the same time, few beneficial effects were noted. There was a change in bone density in those women over 70 and there was an increase in libido, but other effects were less significant. Similarly a study of Rabkin *et al.* (2000) shows that the administration of 200–500 mg of DHEA daily for eight weeks to a group of HIV+ patients in an open label trial leads to a dramatic improvement in mood, libido and fatigue, but no parameter of immune suppression was improved.

Probably the most positive report to come out in the first part of the year 2000 is that of van Vollenhoven (2000) who summarizes a large experience with DHEA in subjects with SLE. Although this report is generally positive, the clinician may find it difficult to understand how DHEA works in these individuals. The studies of Kalimi's group showed that DHEA antagonizes the effects of glucocorticoids, but van Vollenhoven reports that the major effect of DHEA is to spare the amount of exogenous glucocorticoids needed to keep the disease in check.

Some of the disappointments in these human treatment studies may be explained by the dose of DHEA given and the age of the patients in the study. DHEA levels peak around age 30 and fall to very low levels in the elderly. Most studies evaluate supplementing the elderly. Ceresini *et al.* (2000) show that the administration of just 50 mg a day in the morning to older subjects returns serum levels to that of younger subjects. While that is quantitatively satisfying, what does it mean? Will restoring DHEA levels in the elderly accomplish anything if the subject already has endured years of low levels? What dose should be given to younger subjects to prevent degeneration? Should they be given supraphysiologic amounts? Further, are the only important levels those at baseline in the morning? Both DHEA and cortisol are hormones of stress. When does one need DHEA? Is it at baseline on a relaxed day or is it when the individual is stressed and unopposed cortisol causes lasting damage?

CONCLUSION

After all this research how might a clinician view DHEA? Probably with confusion, frustration and maybe even a little bit of hostility. The ready availability of this steroid hormone in the open market in the United States puts enormous pressure on clinicians to explain its uses and dangers to patients. Other steroids like testosterone, estradiol and dexamethasone are not available in such an unregulated manner; they are dispensed only when the physician feels there are bonafide indications and realistic studies to document a positive benefit to risk ratio. Because of a quirk in the United States' law, DHEA is not given such protection and as Baulieu *et al.* (2000) state, "the commercial availability of DHEA outside the regular pharmaceutical-medical network in the United States creates a real pubic health problem...." If a clinician tries to resolve this problem himself the information he can find is confounding, contradictory and poorly established.

The internet is full of such a babble of commercial information, misinformation and overstated claims that it is nearly a disaster for anyone trying to seek a balanced, objective, scientifically based understanding of DHEA. Major textbooks are clearly quiet on potential uses and lead the clinician to think that the area of DHEA research is moribund with regards to potential application. The modern medical literature is clearly more balanced and complete, but as the area of DHEA research is in its infancy, the data is far from understandable from any brief review. Is DHEA a biologically active steroid in its own right or does it really only act as an androgen or estrogen precursor? Does DHEA interact with the gluco-corticoid system? As DHEA decreases with age, is there an advantage of taking supplements and preventing levels from reaching that of the untreated octogen-arian? If so, what dose should be used, physiologic or pharmacologic? At what age should therapy be started? Possibly DHEA is involved in the genesis of Alzheimer's. If so, would those at risk benefit from taking this relatively benign, inexpensive medication? Are there special subgroups who would be harmed by DHEA?

Perhaps, it is because of all these questions that the present volume is needed. It is impossible for a busy clinician to understand the vast area of DHEA research by personally reviewing the literature. There are too many complex intricacies. Would the administration of DHEA benefit some elderly, but harm frail patients? Would the erectile function of some be improved with DHEA, but the adverse effect of the elevated level of androgen cause greater problems with prostatic dis-ease in others? An explication of these questions can only be gained when experts in select, limited fields review all the data in one place and delineate what is known, what is unknown and what the potential side effects of any therapy might be. Hopefully the chapters in this book will serve as a major step in a clinician's understanding of DHEA.

REFERENCES

Baulieu, E.E., Thomas, G., Legrain, S., Lahlou, N., Roger, M., Debuire, B. *et al.* (2000) Dehydroepiandrosterone (DHEA), DHEA sulfate, and aging: contribution of the DHEAge study to a sociobiomedical issue. *Proc. Natl. Acad. Sci. USA*, **97**, 4279–4284.

Brown, R.C., Cascio, C. and Papadopoulos, V. (2000) Pathways of neurosteroid biosynthesis in cell lines from human brain: regulation of dehydroepiandrosterone formation by oxidative stress and beta-armyloid peptide. *Journal of Neurochem.*, **74**, 847–859.

Ceresini, G., Morganti, S., Rebecchi, I., Freddi, M., Ceda, G.P., Banchini, A. *et al.* (2000) Evaluation of the circadian profiles of serum dehydroepiandrosterone (DHEA), cortisol, and cortisol/DHEA molar ratio after a single oral administration of DHEA in elderly subjects. *Metabolism*, **49**, 548–551.

Fitzgerald, P.A. (1997) DHEA is being studied for treatment of Lupus erythematosous. In L.M. Tierney, S.J. McPhee and M.A. Papadakis (eds), *Current medical diagnosis, and treat-ment, 36th Edition*, Appleton and Lange, Stamford, CT, p. 1046.

Goodyer, I.M., Herbert, J., Tamplin, A. and Altham, P.M. (2000) First-episode major depression in adolescents. Affective, cognitive and endocrine characteristics of risk status and predictors of onset. *Brit. J. Psychiat.*, **176**, 142–149.

Hayashi, T., Esaki, T., Muto, E., Kano, H., Asai, Y., Thakur, N.K. *et al.* (2000) Dehydro-epiandrosterone retards atherosclerosis formation through its conversion to estrogen: the possible role of nitric oxide. *Arterioscler. Thromb. Vasc. Biol.*, **20**, 782–792.

Hazzard, W.R., Andres, R., Bierman, E.L. and Blass, J.P. (1990) *Principles of geriatric medicine and gerontology, 2nd edition*. McGraw-Hill Information Services Co., New York.

Hillen, T., Lun, A., Reischies, F.M., Borchelt, M., Steinhagen-Thiessen, E. and Schaub, R.T. (2000) DHEA-S plasma levels and incidence of Alzheimer's disease. *Biol. Psychiat.*, **47**, 161–163.

Hu, Y., Cardounel, A., Gursoy, E., Anderson, P. and Kalimi, M. (2000) Anti-stress effects of dehydroepiandrosterone: protection of rats against repeated immobilization stress-induced weight loss, glucocorticoid receptor production, and lipid peroxidation. *Biochem. Pharmacol.*, **59**, 753–762.

Jarrar, D., Wang, P., Cioffi, W.G., Bland, K.I. and Chaudry, I.H. (2000) Mechanisms of the salutary effects of dehydroepiandrosterone after trauma-hemorrhage: direct or indirect effects on cardiac and hepatocellular functions? *Arch. Surg.*, **135**, 416–422.

Kanik, K.S., Chrousos, G.P., Schumacher, H.R., Crane, M.L., Yarboro, C.H. and Wilder, R.L. (2000) Adrenocorticotropin, glucocorticoid, and androgen secretion in patients with new onset synovitis/rheumatoid arthritis: relations with indices of inflammation. *J. Clin. Endocr. Metab.*, **85**, 1461–1466.

Kiechl, S., Willeit, J., Bonora, E., Schwarz, D. and Xu, Q. (2000) No association between dehydroepiandrosterone sulfate and development of atherosclerosis in a prospective population study (Bruneck Study). *Arterioscler. Thromb. Vasc. Biol.*, **20**, 1094–1100.

Loriaux, D.L. (1996) The adrenal cortex. In J.C. Bennett and F. Plum (eds), *Cecil textbook of medicine, 20th Edition*, W.B. Sanders Co., Philadelphia, PA, pp. 1245–1252.

Moriyama, Y., Yasue, H., Yoshimura, M., Mizuno, Y., Nishiyama, K., Tsunoda, R. *et al.* (2000) The plasma levels of dehydroepiandrosterone sulfate are decreased in patients with chronic heart failure in proportion to the severity. *J. Clin. Endocr. Metab.*, **85**, 1834–1840.

Morrison, M.F., Redei, E., TenHave, T., Parmelee, P., Boyce, A.A., Sinha, P.S. *et al.* (2000) Dehydroepiandrosterone sulfate and psychiatric measures in a frail, elderly residential care population. *Biol. Psychiat.*, **47**, 144–150.

Rabkin, J.G., Ferrando, S.J., Wagner, G.J. and Rabkin, R. (2000) DHEA treatment for HIV+ patients: effects on mood, androgenic and anabolic parameters. *Psychoneuroendocrinology*, **25**, 53–68.

Reiter, W.J., Pycha, A., Schatzl, G., Klingler, H.C., Mark, I., Auterith, A. *et al.* (2000) Serum dehydroepiandrosterone sulfate concentrations in men with erectile dysfunction. *Urology*, **55**, 755–758.

Shi, J., Schulze, S. and Lardy, H.A. (2000) The effect of 7-oxo-DHEA acetate on memory in young and old C57BL/6 mice. *Steroids*, **65**, 124–129.

VanVollenhoven, R.F. (2000) Dehydroepiandrosterone in systemic lupus erythematosus. *Rheum. Dis. Clin. North Am.*, **26**, 349–362.

Williams, G.H. and Dluhy, R.G. (1998) Diseases of the adrenal cortex. In A.S. Fauci, E. Braunwald, D.J. Isselbacker, J.D. Wilson, J.B. Martin, D.L. Kasper, S.L. Hauser and D.L. Longo (eds), *Harrison's principles of internal medicine*, McGraw-Hill, Philadelphia, PA, p. 2035.

Zietz, B., Reber, T., Oertel, M., Gluck, T., Scholmerich, J. and Straub, R.H. (2000) Altered function of the hypothalamic stress axes in patients with moderately active systemic lupus erythematosus. II. Dissociation between androstenedione, cortisol, or dehydroepiandrosterone and interleukin 6 or tumor necrosis factor. *J. Rheumatol.*, **27**, 911–918.

Chapter 13

DHEA replacement and cognition in healthy elderly humans

A summary of results from placebo controlled experiments

Oliver T. Wolf and Clemens Kirschbaum

OVERVIEW

Dehydroepiandrosterone (DHEA) and its sulfated ester (DHEAS) has gained significant interest in the academic community and in the lay public due to claims it might act as an "anti-ageing hormone" or an "endocrine fountain of youth". The strong age associated decline of the steroid in humans and the demonstration of multiple beneficial effects of DHEA in rodents has excited scientists as well as the media. DHEA is available in the United States as an over the counter medication, since it is classified as a food supplement. In addition, DHEA can be ordered via the internet worldwide. Currently an unknown amount of elderly people take DHEA without medical indication or supervision.

The present chapter will describe briefly DHEA/S effects in the rodent brain and on rodent behavior (for an in depth review see Wolf and Kirschbaum, 1999). Thereafter epidemiological and experimental human studies will be summarized. Special emphasis will be placed on results from four studies from our laboratory, which investigated the acute or sub-acute effects of DHEA replacement on cognition and emotion in young and elderly healthy humans. Thereafter recent long-term DHEA trials will be discussed followed by a conclusion on the current state of DHEA replacement in elderly humans.

DHEA EFFECTS IN RODENTS

DHEA/S is the most abundant steroid hormone in man, however its concentrations are very low in laboratory animals (e.g. rats and mice, see Cutler *et al.*, 1978; Fleshner *et al.*, 1997). In addition, only in humans and non-humane primates a pronounced developmental pattern of DHEA/S secretion (with a linear decline during ageing) is observed (Cutler *et al.*, 1978; Lane *et al.*, 1997; Herndon *et al.*, 1999). These species differences are crucial when one discusses potential reasons for the conflicting results obtained with rodents compared to humans.

Action of DHEA in the rodent central nervous system (CNS)

DHEA and DHEAS (together with several other steroids) are produced in the rodent CNS. Therefore, these steroids have been termed "neurosteroids" (see for reviews (Baulieu, 1997; Baulieu and Robel, 1998 and the chapters in this book by Papadopoulos *et al.*, and Barbaccia *et al.*). With respect to primates or humans the empirical evidence for DHEA synthesis in the brain however is sparse.

DHEA/S acts primarily through interactions with neurotransmitter–receptors (see below) on the cell surface. This has been referred to as non-genomic action, which is in contrast to the classic action of steroids which consists of binding to an intracellular receptor (see for review Joels, 1997).

DHEA/S influences several neurotransmitter receptors crucially involved in learning and memory (Wolf and Kirschbaum, 1999). DHEA/S act as negative non-competitive modulators of the $GABA_A$ receptor. GABA is the primary inhibitory transmitter in the CNS (Majewska, 1992, see also her chapter in this book). DHEAS also has effects on sigma receptors, which are functionally linked with the NMDA receptor. Sigma agonistic effects of DHEA have been shown by several laboratories (Monnet *et al.*, 1995; Bergeron *et al.*, 1996; Maurice *et al.*, 1996). DHEA/S also influences several neurotransmitter systems. Effects have been observed on the cholinergic system, which is important for learning (Rhodes *et al.*, 1996) and the hypothalamic serotonergic system, which modulates food intake (Abadie *et al.*, 1993; Svec *et al.*, 1995; Svec and Porter, 1997).

Multiple effects of DHEA/S on neuronal electrophysiology have been observed, most important for the present chapter are effects on models of hippocampal plasticity (Diamond *et al.*, 1996; Yoo *et al.*, 1996). Moreover, DHEAS prevents the reduction in hippocampal plasticity typically observed after stress (Diamond *et al.*, 1999; Kaminiska *et al.*, 2000). The antiglucocorticoid action of DHEA is discussed in more detail in the section covering our human DHEA-stress study.

In addition to the neuro-excitatory effects of DHEA/S summarized above, there is also strong evidence for neuroprotective effects of DHEA/S (see Bologa *et al.*, 1987; Kimonides *et al.*, 1998; Roberts *et al.*, 1987). An important role of DHEA/S in neurodevelopment has also been suggested by the elegant work of Mellon and coworkers (Compagnone *et al.*, 1995; see also Mellon's chapter in this book).

DHEA/S effects on cognition and emotion in rodents

There are now numerous reports on memory enhancing or anti amnestic effects of DHEA/S in young (Flood *et al.*, 1988, 1992; Maurice *et al.*, 1997, 1998; Melchior and Ritzmann, 1996; Reddy and Kulkarni, 1998a; Urani *et al.*, 1998) and old (Flood and Roberts, 1988; Reddy and Kulkarni, 1998b) mice. Data are less consistent in rats (Fleshner *et al.*, 1997; Frye and Sturgis, 1999). DHEA/S enhances emotional as well as spatial memory in mice. Most studies, which tested multiple doses report an inverted U shaped dose response curve, however the effective doses vary drastically between the studies. In addition to its effect on memory, DHEAS also has anxiolytic properties (Melchior and Ritzmann, 1994; Prasad *et al.*, 1997; Reddy *et al.*, 1998) which of course might indirectly influence performance in memory tasks.

Taken together, multiple beneficial effects of DHEA/S have been observed in the rodent brain and these effects translate into memory enhancing and anti-depressant action. The human studies reported in the next section therefore were based on convincing basic science data.

DHEA AND COGNITION OR MOOD IN HUMANS

Since DHEA and DHEAS concentrations decrease so dramatically with age in humans and multiple beneficial effects of DHEA in the CNS have been observed in rodents, several studies were conducted to investigate the relationship between DHEA and CNS functions in healthy elderly humans and elderly patients with psychiatric disorders. This section will first summarize studies which investigated associations between DHEA and measures of cognition or emotions. The second section discusses data from experimental trials.

Epidemiological studies

Many studies exist which tried to relate DHEAS levels and global and/or cognitive functioning in elderly humans in a cross sectional or longitudinal design (see for review Wolf and Kirschbaum, 1999). Some studies report an association between low DHEAS levels and impairments in global measures of functional abilities (e.g. activities of daily living) (Rudman *et al.*, 1990; Berkman *et al.*, 1993; Ravaglia *et al.*, 1996). Moreover, lower DHEAS levels might be associated with depressive symptoms in normal elderlies (Barret Connor *et al.*, 1999; Berr *et al.*, 1996). However a relationship to cognitive test performance could not be observed in large cross-sectional (Berr *et al.*, 1996) or longitudinal (Barrett-Connor and Edelstein, 1994; Moffat *et al.*, 2000; Kalmijn *et al.*, 1998; Yaffe *et al.*, 1998) studies. Moreover preliminary reports regarding reduced DHEAS levels in patients with Alzheimer's disease could not be replicated by others (Wolf and Kirschbaum, 1999). Taken together these studies demonstrate that low DHEAS levels seem to be associated with poorer health, impaired global functioning, or psychological well being, while there seems to be no relationship to cognition.

Experimental studies in humans

As outlined in the previous sections, results from studies investigating associations between DHEA/S levels and cognition or well-being are conflicting. Only experimental trials in which the steroid is given to volunteers in a double blind fashion are able to answer the question, whether DHEA/S effects in humans might be as strong as they are in rodents. We performed four experimental trials which measured cognitive performance and mood after DHEA administration. These will be discussed in this section.

Study I: DHEA has no acute effects on cognition or mood in young men (Wolf et al., 1997a)

As a first step we tested, the effects of a single high DHEA dose (300 mg) on memory and mood in 36 young men (mean age = 25 years.). The subjects were tested late in the evening, at a time when endogenous DHEA levels are low and when cognitive performance is relatively poor. In this study a mood questionnaire and four memory tests were employed. On an initial baseline test-day every subject received placebo. On the second test day half of the subjects received DHEA, while the other half again received placebo. DHEA treatment did neither alter mood nor did it influence performance in any of the memory tests. These results were therefore in contrast to the rodent literature suggesting memory enhancing effects of a single DHEA treatment in young mice and rats.

Study II: two weeks of DHEA replacement does not enhance cognition or mood in healthy elderly men and women (Wolf et al., 1997b)

In a second study, the cognitive and psychological effects of a two week DHEA replacement (50 mg/day) in healthy elderly women (n = 15) and men (n = 25) with a mean age of 70 years were tested. A large cognitive test battery measuring attention, response inhibition, verbal as well as visual memory was employed. Moreover, several mood, well-being and depression questionnaires were used. As expected the elderly subjects had low DHEA and DHEAS levels, similar to those previously reported by others. DHEA and DHEAS levels increased after the treatment and reached levels typically observed in young individuals. DHEA replacement however had no effects on cognitive test performance. It tended to increase mood in women, but not in men. Further emphasizing these, negative results was the observation that the subjects were unable to correctly guess which treatment they had received. These findings, which were the first to test the effects of DHEA replacement on cognitive performance using standardized tests in a double blind study design therefore did not support the hypothesis, that DHEA replacement has strong beneficial effects on cognition or mood in healthy elderly subjects.

Study III: DHEA has electophysiological effects in elderly men, which do not translate into changes in cognitive test performance (Wolf et al., 1998b)

A third experiment used event related potentials (ERPs) derived from EEG recordings in order to investigate possible CNS effects of the steroid treatment. Event related potentials are a sensitive measure of information processing in the CNS and can be a valuable tool in neuro-endocrine research. In this study, the so called P 300 (an electrophysiological index of information processing in short term memory, which shows marked alterations with ageing (Polich, 1996)) was investigated using a standard auditory oddball paradigm. In addition visual, spatial and semantic memory tests were used. The subjects were 14 healthy elderly men with a mean age of 71 years. DHEA dose and treatment length were similar to Experiment II (50 mg/day for two weeks). The EEG results indicated that the P 300 amplitude was enhanced after DHEA treatment, while no change in the P 300 latency was observed. Again no effect on memory test performance could be observed. This finding suggests that with more sensitive measurement techniques subtle effects of DHEA replacement on CNS stimulus processing can be observed, which do not seem to be strong enough to influence test performance or measures of well-being.

Study IV: DHEA has opposing effects on memory and attention after psychosocial stress (Wolf et al., 1998a)

The fourth study set out to investigate whether the hypothesized antiglucocorticoid action of DHEA (Kalimi et al., 1994) would protect hippocampal mediated declarative memory from the impairing effects of psychosocial stress (Kirschbaum et al., 1996; Lupien et al., 1997). Seventy five elderly men and women (mean age 67 years), participated in this experiment and received DHEA (50 mg/day) or placebo in a double blind control group design (Wolf et al., 1998a). Memory and attention was tested before and after exposure to a standardized psychosocial laboratory stressor (Trier Social Stress Test; TSST; Kirschbaum et al., 1993). This stressor consists of public speaking (5 min) followed by mental arithmetic (5 min) in front of a committee and has been shown to consistently increase cortisol levels in numerous previous studies from our laboratory. The endocrine results showed that DHEA-replaced subjects tended to show an increased free cortisol response with DHEA treatment (Kudielka et al., 1998; Wolf et al., 1998b). After stress subjects under DHEA recalled less items from the "picture memory test", which they had learned before stress exposure (hippocampal mediated declarative memory). However they performed better in an attention task than subjects from the placebo group. This finding might be mediated by the enhanced free cortisol response of the DHEA replaced subjects. Exogenous cortisol administration impairs declarative memory, with recall being most sensitive (de Quervain et al., 2000), but data from a study employing questionnaires suggest that cortisol enhances attention (Plihal et al., 1996). Our findings document a complex DHEA by stress by cognitive domain interaction with impairing as well as protecting properties of the steroid. However the results do not support the idea that DHEA replacement might protect hippocampal dependent declarative memory from the impairing effects of psychosocial stress. These results are similar to the report of a DHEA induced impairment of hippocampal mediated fear conditioning in rats (Fleshner et al., 1997), but seem to be in contrast to electrophysiological studies, which have demonstrated, that DHEAS protects hippocampal plasticity from the impairing effects of glucocorticoids (Diamond et al., 1999; Kaminska et al., 2000).

Long-term studies from other laboratories

Morales et al. (1994) reported that three months of DHEA replacement (50 mg/day) in 30 age advanced subjects (mean age 54) increased the "sense of well being" in 82% of women and 67% of men in a placebo controlled double blind experiment. Unfortunately well being was only assessed with an open interview and no information is provided how these qualitative data were transformed into the dichotomous variable used for statistical analysis. This initial positive study was one key factor in the creation of the "DHEA-hype" by the media in the next few years. Flynn and coworkers performed a double blind placebo controlled crossover study with 39 elderly men who received 100 mg/day DHEA or placebo over a period of three months (Flynn et al., 1999). No changes in the satisfaction to perform activities of daily living as assessed with a questionnaire were detected. Arlt et al. (1999) had previously reported that DHEA improved mood and sexuality (as assessed with questionnaires) in women with adrenal insufficiency at a mean age of 42 years. However the authors failed to observe similar beneficial effects of DHEA (50 mg/day over four months) in elderly men (n = 22; mean age = 59 years), which were screened for low endogenous DHEAS levels before study entry (Arlt *et al.* 2000). Another study (Barnhardt *et al.*, 1999) tested the effects of DHEA replacement (50 mg/day for three months) in 60 peri-menopausal women (n = 60, mean age = 50 years) with mood complaints but could not detect any beneficial effects of DHEA. In contrast, positive DHEA effects have been reported in age advanced patients with depression. Two recent double blind trials observed that DHEA treatment had antidepressant effects (Bloch *et al.*, 1999; Wolkowitz *et al.*, 1999; see the chapter by Reus and Wolkowitz in this book).

In summary, human experimental studies on the effects of DHEA replacement on cognition in young or elderly healthy people have consistently failed to observe any beneficial effects. While the postulated mood enhancing properties might occur in depressed patients, but these effects have not been observed in several placebo controlled studies in normal healthy elderly individuals so far.

DHEA vs estrogen and testosterone replacement

When discussing possible effects of DHEA replacement it seems to be appropriate to compare its reported effects (or the absence of them) with the effects of the two primary human sex steroids estradiol and testosterone. The studies by Sherwin and colleagues suggest that estradiol enhances verbal memory performance (Phillips and Sherwin, 1992; Sherwin and Tulandi, 1996) and we could recently demonstrate that this effect occur in less than two weeks in women who had been postmenopausal for more than 10 years (Wolf *et al.*, 1999). Another recent study observed even broader effects on memory and reasoning (Duka, Tasker and McGowan, 2000). Other studies reported enhanced attention after estradiol treatment or failed to find any positive effect. These conflicting results might be caused by the variability in cognitive tests and/or different estradiol compounds used (see for recent reviews, Haskell *et al.*, 1997; Rice *et al.*, 1997; Sherwin, 1997). In addition to its effects on memory, anti-depressant effects of estradiol replacement have been documented in several studies (Zweifel and O'Brien, 1997). Possible underlying CNS mechanisms of these beneficial effects of estradiol are summarized by McEwen and Alves (1999).

There are only a few testosterone replacement studies looking at cognitive outcome measures. Cognitive enhancement has been found in two studies (Cherrier *et al.*, 1998; Janowsky *et al.*, 1994), while two other studies failed to find beneficial effects (Sih *et al.*, 1997; Wolf *et al.*, 2000). At least mood and libido enhancement have been documented repeatedly (Sternbach, 1998).

Several beneficial effects of estrogen replacement in postmenopausal women are now demonstrated, while testosterone replacement in elderly men is still a neglected research area. Therefore estrogen as a cognition and/or mood enhancer for postmenopausal women is much better backed up by scientific data than DHEA. A comparison between DHEA and testosterone replacement efficacy for elderly men is not possible at the moment, due to the lack of appropriate experimental studies regarding both hormones.

SUMMARY AND CONCLUSION

The present chapter briefly summarized the broad and remarkable range of beneficial effects of DHEA/S treatment in rodents and compared those effects to results obtained in experimental studies in humans. This comparison revealed profound differences in the methods used as well as in the results obtained.

First, on a more general note, DHEA concentrations are low in rodents compared to humans. This raises the question whether rodents are adequate animal models to study DHEA effects. Extrapolation from data obtained in rodents to humans should be made with caution in particular with respect to DHEA. Second, most of the animal studies have investigated the effects of rather acute DHEA/S treatments, often with relatively high DHEA/S doses. DHEA replacement in elderly humans in contrast would consist of a long-term treatment with relatively low DHEA doses.

Today, the human research in this field is still relatively sparse, so that several areas of potentially beneficial long-term DHEA treatment have not been investigated. Beneficial effects on electrophysiological indices of CNS functioning have been obtained after a two week DHEA replacement in elderly men (Wolf *et al.*, 1998b), however studies looking at cognition or mood have failed to find effects of DHEA in healthy subjects (see above). These studies document, that the fast neuroactive effects of DHEA as observed in rodents can be detected with sensitive electrophysiological methods, but does not seem to be strong enough to improve cognition in healthy young or elderly adults. However prolonged DHEA treatment has been reported to reduce depressive symptoms in psychiatric patients or patients with adrenal insufficiency. The time course of these effects and possible mediating variables should be investigated in the future.

Taken together the multiple beneficial effects of DHEA/S as observed in rodents could, as of today, not be substantiated by similar effects in healthy elderly humans. Of course, it is much too early to dismiss the possibility of positive effects of DHEA replacement on human brain ageing. Future studies are needed to identify clinical patients who might show favorable responses to DHEA/S treatment.

ACKNOWLEDGMENTS

This work was supported by DFG grants Ki 537/6-1, He1013/13-1, Wo 733/1-1 and Wo 733/2-1.

REFERENCES

Abadie, J.M., Wright, B., Correa, G., Browne, E.S., Porter, J.R. and Svec, F. (1993) Effect of dehydroepiandrosterone on neurotransmitter levels and appetite regulation of the obese Zucker rat. The obesity research program. *Diabetes*, **42**, 662–669.

Arlt, W., Callies, F., van Vlijmen, J.C., Koehler, I., Reincke, M., Bidlingmaier, M. *et al.* (1999) Dehydroepiandrosterone replacement in women with adrenal insufficiency. *New Engl. J. Med.*, **341**, 1013–1020.

Arlt, W., Callies, F., Koehler, I., van Vlijmen, J.C., Huebler, D., Oettel, M. *et al.* (2000) Dehydroepiandrosterone (DHEA) supplementation in elderly men with low endogenous serum DHEAS. *Exp. Clin. Endocr. Diab.*, **108**, S195.

Barnhart, K.T., Freeman, E., Grisso, J.A., Rader, D.J., Sammel, M., Kapoor, S. *et al.* (1999) The effect of dehydroepiandrosterone supplementation to symptomatic perimenopausal women on serum endocrine profiles, lipid parameters, and health-related quality of life. *J. Clin. Endocrinol. Metab.*, **84**, 3896–3902.

Barrett-Connor, E. and Edelstein, S.L. (1994) A prospective study of dehydroepiandrosterone sulfate and cognitive function in an older population: the Rancho Bernardo study. *J. Am. Geriat. Soc.*, **42**, 420–423.

Barrett-Connor, E., von Muhlen, D., Laughlin, G.A. and Kripke, A. (1999) Endogenous levels of dehydroepiandrosterone sulfate, but not other sex hormones, are associated with depressed mood in older women: the Rancho Bernardo study. *J. Am. Geriat. Soc.*, **47**, 685–691.

Baulieu, E.E. (1997) Neurosteroids: of the nervous system, by the nervous system, for the nervous system. *Recent Prog. Horm. Res.*, **52**, 1–32.

Baulieu, E.E. and Robel, P. (1998) Dehydroepiandrosterone (DHEA) and dehydro-epiandrosterone sulfate (DHEAS) as neuroactive neurosteroids. *Proc. Natl. Acad. Sci. USA*, **95**, 4089–4091.

Bergeron, R., de Montigny, C. and Debonnel, G. (1996) Potentiation of neuronal NMDA response induced by dehydroepiandrosterone and its suppression by progesterone: effects mediated via sigma receptors. *J. Neurosci.*, **16**, 1193–1202.

Berkman, L.F., Seeman, T.E., Albert, M., Blazer, D., Kahn, R., Mohs, R. *et al.* (1993) High, usual and impaired functioning in community-dwelling older men and women: findings from the MacArthur foundation research network on successful aging. *J. Clin. Epidemiol.*, **46**, 1129–1140.

Berr, C., Lafont, S., Debuire, B., Dartigues, J.F. and Baulieu, E.E. (1996) Relationships of dehydroepiandrosterone sulfate in the elderly with functional, psychological, and mental status, and short-term mortality: a French community-based study. *Proc. Natl. Acad. Sci. USA*, **93**, 13410–13415.

Bloch, M., Schmidt, P.J., Danaceau, M.A., Adams, L.F. and Rubinow, D.R. (1999) Dehydro-epiandrosterone treatment of midlife dysthymia. *Biol. Psychiat.*, **45**, 1533–1541.

Bologa, L., Sharma, J. and Roberts, E. (1987) Dehydroepiandrosterone and its sulfated derivative reduce neuronal death and enhance astrocytic differentiation in brain cell cultures. *J. Neurosci. Res.*, **17**, 225–234.

Cherrier, M., Asthana, S., Plymate, S., Baker, L., Matsumoto, A., Bremner, W. *et al.* (1998) Effects of testosterone on cognition in healthy elderly men. *Soc. Neurosci. (Abstracts)*, **24**, 24118.

Compagnone, N.A., Bulfone, A., Rubenstein, J.L. and Mellon, S.H. (1995) Expression of the steroidogenic enzyme P450scc in the central and peripheral nervous systems during rodent embryogenesis. *Endocrinology*, **136**, 2689–2696.

Cutler, Jr. G.B., Glenn, M., Bush, M., Hodgen, G.D., Graham, C.E. and Loriaux, D.L. (1978) Adrenarche: a survey of rodents, domestic animals, and primates. *Endocrinology*, **103**, 2112–2118.

de Quervain, D.J., Roozendaal, B., Nitsch, R.M., McGaugh, J.L. and Hock, C. (2000) Acute cortisone administration impairs retrieval of long-term declarative memory in humans. *Nature Neurosci.*, **3**, 313–314.

Diamond, D.M., Branch, B.J. and Fleschner, M. (1996) The neurosteroid dehydroepi-androsterone sulfate (DHEAS) enhances hippocampal primed burst, but not long-term, potentiation. *Neurosci. Lett.*, **202**, 204–208.

Diamond, D.M., Fleshner, M. and Rose, G.M. (1999) The DHEAS-induced enhancement of hippocampal primed burst potentiation is blocked by psychological stress. *Stress*, **3**, 107–121.

Fleshner, M., Pugh, C.R., Tremblay, D. and Rudy, J.W. (1997) DHEA-S selectively impairs contextual-fear conditioning: support for the antiglucocorticoid hypothesis. *Behav. Neurosci.*, **111**, 512–517.

Flood, J.F., Morley, J.E. and Roberts, E. (1992) Memory-enhancing effects in male mice of pregnenolone and steroids metabolically derived from it. *Proc. Natl. Acad. Sci. USA*, **89**, 1567–1571.

Flood, J.F. and Roberts, E. (1988) Dehydroepiandrosterone sulfate improves memory in aging mice. *Brain Res.*, **448**, 178–181.

Flood, J.F., Smith, G.E. and Roberts, E. (1988) Dehydroepiandrosterone and its sulfate enhance memory retention in mice. *Brain Res.*, **447**, 269–278.

Flynn, M.A., Weaver-Osterholtz, D., Sharpe-Timms, K.L., Allen, S. and Krause, G. (1999) Dehydroepiandrosterone replacement in aging humans. *J. Clin. Endocr. Metab.*, **84**, 1527–1533.

Frye, C.A. and Sturgis, J.D. (1995) Neurosteroids affect spatial/reference, working, and long-term memory of female rats. *Neurobiol. Learn. Mem.*, **64**, 83–96.

Haskell, S.G., Richardson, E.D. and Horwitz, R.I. (1997) The effect of estrogen replacement therapy on cognitive function in women: a critical review of the literature. *J. Clin. Epidemiol.*, **50**, 1249–1264.

Herndon, J.G., Lacreuse, A., Ladinsky, E., Killiany, R.J., Rosene, D.L. and Moss, M.B. (1999) Age-related decline in DHEAS is not related to cognitive impairment in aged monkeys. *Neuroreport*, **10**, 3507–3511.

Janowsky, J.S., Oviatt, S.K. and Orwoll, E.S. (1994) Testosterone influences spatial cognition in older men. *Behav. Neurosci.*, **108**, 325–332.

Joels, M. (1997) Steroid hormones and excitability in the mammalian brain. *Front Neuroendocrinol.*, **18**, 2–48.

Kalimi, M., Shafagoj, Y., Loria, R., Padgett, D. and Regelson, W. (1994) Anti-glucocorticoid effects of dehydroepiandrosterone (DHEA). *Mol. Cell. Biochem.*, **131**, 99–104.

Kalmijn, S., Launer, L.J., Stolk, R.P., de Jong, F.H., Pols, H.A., Hofman, A. *et al.* (1998) A prospective study on cortisol, dehydroepiandrosterone sulfate, and cognitive function in the elderly. *J. Clin. Endocr. Metab.*, **83**, 3487–3492.

Kaminska, M., Harris, J., Gijsbers, K. and Dubrovsky, B. (2000) Dehydroepiandrosterone sulfate (DHEAS) counteracts decremental effects of corticosterone on dentate gyrus LTP. Implications for depression. *Brain Res. Bull.*, **52**, 229–234.

Kimonides, V.G., Khatibi, N.H., Svendsen, C.N., Sofroniew, M.V. and Herbert, J. (1998) Dehydroepiandrosterone (DHEA) and DHEA-sulfate (DHEAS) protect hippocampal neurons against excitatory amino acid-induced neurotoxicity. *Proc. Natl. Acad. Sci. USA*, **95**, 1852–1857.

Kirschbaum, C., Pirke, K.M. and Hellhammer, D.H. (1993) The "trier social stress test" – a tool for investigating psychobiological stress responses in a laboratory setting. *Neuropsychobiology*, **28**, 76–81.

Kirschbaum, C., Wolf, O.T., May, M., Wippich, W. and Hellhammer, D.H. (1996) Stress- and treatment-induced elevations of cortisol levels associated with impaired declarative memory in healthy adults. *Life Sci.*, **58**, 1475–1483.

Kudielka, B.M., Hellhammer, J., Hellhammer, D.H., Wolf, O.T., Pirke, K.M., Varadi, E. *et al.* (1998) Sex differences in endocrine and psychological responses to psychosocial stress in healthy elderly subjects and the impact of a 2-week dehydroepiandrosterone treatment. *J. Clin. Endocr. Metab.*, **83**, 1756–1761.

Lane, M.A., Ingram, D.K., Ball, S.S. and Roth, G.S. (1997) Dehydroepiandrosterone sulfate: a biomarker of primate aging slowed by calorie restriction. *J. Clin. Endocr. Metab.*, **82**, 2093–2096.

Lupien, S.J., Gaudreau, S., Tchiteya, B.M., Maheu, F., Sharma, S., Nair, N.P.V. *et al.* (1997) Stress-induced declarative memory impairment in healthy elderly subjects: relationship to cortisol reactivity. *J. Clin. Endocr. Metab.*, **82**, 2070–2075.

Majewska, M.D. (1992) Neurosteroids: endogenous bimodal modulators of the GABA-A receptor. Mechanism of action and physiological significance. *Prog. Neurobiol.*, **38**, 379–395.

Maurice, T., Junien, J.L. and Privat, A. (1997) Dehydroepiandrosterone sulfate attenuates dizocilpine-induced learning impairment in mice via sigma 1-receptors. *Behav. Brain Res.*, **83**, 159–164.

Maurice, T., Roman, F.J. and Privat, A. (1996) Modulation by neurosteroids of the *in vivo* (+)-[3H]SKF-10,047 binding to sigma 1 receptors in the mouse forebrain. *J. Neurosci. Res.*, **46**, 734–743.

Maurice, T., Su, T.P. and Privat, A. (1998) Sigma1 (sigma 1) receptor agonists and neurosteroids attenuate B25–35-amyloid peptide-induced amnesia in mice through a common mechanism. *Neuroscience*, **83**, 413–428.

McEwen, B.S. and Alves, S.E. (1999) Estrogen action in the central nervous system. *Endocr. Rev.*, **20**, 279–307.

Melchior, C.L. and Ritzmann, R.F. (1994) Dehydroepiandrosterone is an anxiolytic in mice on the plus maze. *Pharmacol. Biochem. Behav.*, **47**, 437–441.

Melchior, C.L. and Ritzmann, R.F. (1996) Neurosteroids block the memory-impairing effects of ethanol in mice. *Pharmacol. Biochem. Behav.*, **53**, 51–56.

Moffat, S.D., Zonderman, A.B., Harman, S.M., Blackman, M.R., Kawas, C. and Resnick, S.M. (2000) The relationship between longitudinal declines in dehydroepiandrosterone sulfate concentrations and cognitive performance in older men. *Arch. Intern. Med.*, **160**, 2193–2198.

Monnet, F.P., Mahe, V., Robel, P. and Baulieu, E.E. (1995) Neurosteroids, via sigma receptors, modulate the [3H]norepinephrine release evoked by N-methyl-D-aspartate in the rat hippocampus. *Proc. Natl. Acad. Sci. USA*, **92**, 3774–3778.

Morales, A.J., Nolan, J.J., Nelson, J.C. and Yen, S.S.C. (1994) Effects of replacement dose of dehydroepiandrosterone in men and women of advancing age. *J. Clin. Endocr. Metab.*, **78**, 1360–1367.

Phillips, S.M. and Sherwin, B.B. (1992) Effects of estrogen on memory function in surgically menopausal women. *Psychoneuroendocrinology*, **17**, 485–495.

Plihal, W., Krug, R., Pietrowsky, R., Fehm, H.L. and Born, J. (1996) Corticosteroid receptor mediated effects on mood in humans. *Psychoneuroendocrinology*, **21**, 515–523.

Polich, J. (1996) Meta-analysis of P300 normative aging studies. *Psychophysiology*, **33**, 334–353.

Prasad, A., Imamura, M. and Prasad, C. (1997) Dehydroepiandrosterone decreases behavioral despair in high- but not low-anxiety rats. *Physiol. Behav.*, **62**, 1053–1057.

Ravaglia, G., Forti, P., Maioli, F., Boschi, F., Bernardi, M., Pratelli, L. *et al.* (1996) The relationship of dehydroepiandrosterone sulfate (DHEAS) to endocrine-metabolic parameters and functional status in the oldest-old. Results from an Italian study on healthy free-living over-ninety-year-olds. *J. Clin. Endocr. Metab.*, **81**, 1173–1178.

Reddy, D.S., Kaur, G. and Kulkarni, S.K. (1998) Sigma (sigma 1) receptor mediated antidepressant-like effects of neurosteroids in the Porsolt forced swim test. *Neuroreport*, **9**, 3069–3073.

Reddy, D.S. and Kulkarni, S.K. (1998a) The effects of neurosteroids on acquisition and retention of a modified passive-avoidance learning task in mice. *Brain Res.*, **791**, 108–116.

Reddy, D.S. and Kulkarni, S.K. (1998b) Possible role of nitric oxide in the nootropic and antiamnesic effects of neurosteroids on aging- and dizocilpine-induced learning impairment. *Brain Res.*, **799**, 215–229.

Rhodes, M.E., Li, P.K., Flood, J.F. and Johnson, D.A. (1996) Enhancement of hippocampal acetylcholine release by the neurosteroid dehydroepiandrosterone: an *in vivo* microdialysis study. *Brain Res.*, **733**, 284–286.

Rice, M.S., Graves, A.B., McCurry, S.M. and Larson, E.B. (1997) Estrogen replacement therapy and cognitive function in postmenopausal women without dementia. *Am. J. Med.*, **103**, 26S–35S.

Roberts, E., Bologa, L., Flood, J.F. and Smith, G.E. (1987) Effects of dehydroepiandro-sterone and its sulfate on brain tissue in culture and on memory in mice. *Brain Res.*, **406**, 357–362.

Rudman, D., Shetty, K.R. and Mattson, D.E. (1990) Plasma dehydroepiandrosterone sulfate in nursing home men. *J. Am. Geriatr. Soc.*, **38**, 421–427.

Sherwin, B.B. (1997) Estrogen effects on cognition in menopausal women. *Neurology*, **48**, S21–S26.

Sherwin, B.B. and Tulandi, T. (1996) "Add-back"estrogen reverses cognitive deficits induced by a gonadotropin-releasing hormone agonist in women with leiomyomata uteri. *J. Clin. Endocr. Metab.*, **81**, 2545–2549.

Sih, R., Morley, J.E., Kaiser, F.E., Perry, H.M.R., Patrick, P. and Ross, C. (1997) Testo-sterone replacement in older hypogonadal men: a 12-month randomized controlled trial. *J. Clin. Endocr. Metab.*, **82**, 1661–1667.

Sternbach, H. (1998) Age-associated testosterone decline in men: clinical issues for psychiatry. *Am. J. Psychiat.*, **155**, 1310–1318.

Svec, F., Hilton, C.W., Wright, B., Browne, E. and Porter, J.R. (1995) The effect of DHEA given chronically to Zucker rats. *Proc. Soc. Exp. Biol. Med.*, **209**, 92–97.

Svec, F. and Porter, J. (1997) The effect of dehydroepiandrosterone (DHEA) on Zucker rat food selection and hypothalamic neurotransmitters. *Psychoneuroendocrinology*, **22**, S57–S62.

Urani, A., Privat, A. and Maurice, T. (1998) The modulation by neurosteroids of the scopo-lamine-induced learning impairment in mice involves an interaction with sigma1 (sigma 1) receptors. *Brain Res.*, **799**, 64–77.

Wolf, O.T., Naumann, E., Hellhammer, D.H. and Kirschbaum, C. (1998b) Effects of deyhdro-epiandrosterone (DHEA) replacement in elderly men on event related potentials (ERPs), memory and well-being. *J. Gerontol.*, **53**, M385–M390.

Wolf, O.T., Neumann, O., Hellhammer, D.H., Geiben, A.C., Strasburger, C.J., Dressendorfer, R.A. *et al.* (1997b) Effects of a two-week physiological dehydroepiandrosterone substitution on cognitive performance and well-being in healthy elderly women and men. *J. Clin. Endocr. Metab.*, **82**, 2363–2367.

Wolf, O.T. and Kirschbaum, C. (1999) Actions of dehydroepiandrosterone and its sulfate in the central nervous system: effects on cognition and emotion in animals and humans. *Brain Res. Rev.*, **30**, 264–288.

Wolf, O.T., Koster, B., Kirschbaum, C., Pietrowsky, R., Kern, W., Hellhammer, D.H. *et al.* (1997a) A single administration of dehydroepiandrosterone does not enhance memory performance in young healthy adults, but immediately reduces cortisol levels. *Biol. Psychiat.*, **42**, 845–848.

Wolf, O.T., Kudielka, B.M., Hellhammer, J., Hellhammer, D.H. and Kirschbaum, C. (1998a) Opposing effects of DHEA replacement in elderly subjects on declarative memory and attention after exposure to a laboratory stressor. *Psychoneuroendocrinology*, **23**, 617–629.

Wolf, O.T., Kudielka, B.M., Hellhammer, D.H., Törber, S., McEwen, B.S. and Kirschbaum, C. (1999) Two weeks of transdermal estradiol treatment in postmenopausal women and its effect on memory and mood: verbal memory changes are associated with the treatment induced estradiol levels. *Psychoneuroendocrinology*, **24**, 727–741.

Wolf, O.T., Preut, R., Hellhammer, D.H., Kudielka, B.M., Schürmeyer, T.H. and Kirsch-baum, C. (2000) Testosterone and cognition in elderly men: a single testosterone injection impairs verbal fluency, but has no effects on spatial or verbal memory. *Biol. Psychiat.*, **47**, 650–654.

Wolkowitz, O.M., Reus, V.I., Keebler, A., Nelson, N., Friedland, M., Brizendine, L. *et al.* (1999) Double-blind treatment of major depression with dehydroepiandrosterone (DHEA). *Am. J. Psychiat.*, **156**, 646–649.

Yaffe, K., Ettinger, B., Pressman, A., Seeley, D., Whooley, M., Schaefer, C. *et al.* (1998) Neuropsychiatric function and dehydroepiandrosterone sulfate in elderly women: a prospective study. *Biol. Psychiat.*, **43**, 694–700.

Yoo, A., Harris, J. and Dubrovsky, B. (1996) Dose–response study of dehydroepiandrosterone sulfate on dentate gyrus long-term potentiation. *Exp. Neurol.*, **137**, 151–156.

Zweifel, J.E. and O'Brien, W.H. (1997) A meta-analysis of the effect of hormone replacement therapy upon depressed mood. *Psychoneuroendocrinology*, **22**, 189–212.

Index

ACTH *see* adrenocorticotropic hormone
acyl-CoA oxidase 90
adenomas 89
adenylate cyclase; ascorbic acid, 10–11;
  inhibition, 10–11
AD *see* Alzheimer's disease
ADIOL *see* 5-androstene-3β,17β-diol
adipose tissue 118, 167
adrenalectomy 7, 42, 48, 50, 132
adrenal cortex 130, 132, 148, 167, 177
adrenocorticotropic hormone (ACTH)
  6–11, 41, 69, 130, 132, 134, 148, 173,
  174, 182
adrenodoxin 103
adrenodoxin reductase functions 103
ageing; dehydroepiandrosterone and
  sulfate decline of levels 24–6, 33, 54–5,
  58, 71, 140, 155, 156; overview 40, 54,
  56, 140, 155, 156, 176–85; plasma
  levels 26, 54, 55, 179; in Alzheimer's
  disease 54, 56, 179, 183; in healthy
  eldrely population 187, 189–93;
  7-hydroxy-DHEA levels 28–32
AIDS 179, 180
alcohol *see* ethanol
allopregnanolone *see* 3α-hydroxy-
  5α-pregnane-20-one
allotetrahydroxydeoxycorticosterone
  see 3α,5α-tetrahydroxydeoxycorticosterone
alphaxalone 65
alternative pathway 135, 137, 138
Alzheimer's disease (AD) 54, 56, 59, 71,
  137–41
γ-aminobutyric acid type A receptor
  (GABA$_A$); benzodiazepine action
  regulation 66, 70, 134, 158, 159; binding
  sites for steroids 66, 67, 101, 102; brain
  localization and steroid effects 25, 65, 71,
  106, 110, 134, 142, 188; central
  depressive effects of steroids 13, 158;
  learning and memory role, neurosteroid
  interactions 70, 111; ligand diversity

101, 102; progesterone effects 15, 16;
  steroid modulation 15–17, 59, 65–8,
  129
alprazolam 158
amphophilic cell 89
amphophilic cell foci (APF) 89–94
amygdala 107
β-amyloid peptide 51, 137, 138, 140, 141
androgens 25, 42, 68, 69, 72, 81, 88, 130,
  132, 142
5α-androstane-3α,17β-diol 14–17
5α-androstane-3β,17β-diol 82, 117, 120–2
androstenedione *see* 4-androstene-3,17-
  dione
5-androstene-3β,17β-diol (ADIOL) 48, 82,
  106, 117
4-androstene-3,17-dione 25, 104, 167, 168
5-androstene-3β,7α,17β-triol 82
androsterone 66, 67, 69, 83
animal models 190, 193
AOX *see* acyl-CoA oxidase
APF *see* amphophilic cell foci
appetite modulation 147, 149–51
anticarcinogenic effects 93
antiglucocorticoid 24, 27, 48, 49, 52, 81,
  84, 122, 188, 191
antiobesity 83, 84, 86
ascorbic acid 4–11
atherosclerosis 181
astrocytes 27, 107, 136–8, 141
anxiety 15, 16
ATP 85
autocrine 28
autoimmune diseases 34
axonal outgrowth 109, 110

basophilic cell foci (BCF) 90, 93
BCF *see* basophilic cell foci
benzodiazepine receptor 1
bile acids 81, 120
blood glucose 84
blood–brain barrier 25, 45–6, 71, 129

Printed and bound by CPI Group (UK) Ltd, Croydon, CR0 4YY

23/10/2024

01778226-0002